Fundamentals of Horticulture

Fundamentals of Horticulture

Beatrice Pollard

Larsen & Keller
www.larsen-keller.com

Fundamentals of Horticulture
Beatrice Pollard
ISBN: 978-1-64172-605-4 (Hardback)

Larsen & Keller
Published by Larsen and Keller Education,
5 Penn Plaza,
19th Floor,
New York, NY 10001, USA

Cataloging-in-Publication Data

Fundamentals of horticulture / Beatrice Pollard.
p. cm.
Includes bibliographical references and index.
ISBN 978-1-64172-605-4
1. Horticulture. 2. Gardening. 3. Agriculture. I. Pollard, Beatrice.
SB318 .F86 2022
635--dc23

TABLE OF CONTENTS

PREFACE

It is with great pleasure that I present this book. It has been carefully written after numerous discussions with my peers and other practitioners of the field. I would like to take this opportunity to thank my family and friends who have been extremely supporting at every step in my life.

Horticulture refers to the agriculture of plants for various purposes such as food, materials, beauty and comfort. It also includes the growing of vegetables, flowers and fruits for ornamental purposes. Horticulture involves landscape restoration, conservation of plants and soil management. It also deals with landscape and garden construction, design and maintenance as well as arboriculture. Some of the major focus areas within this field are turf management, floriculture, olericulture and pomology. Turf management deals with the production and management of turf grass for various purposes. Floriculture is involved in the production and marketing of floral crops. The production and marketing of vegetables and fruits are termed as olericulture and pomology, respectively. This textbook presents the complex subject of horticulture in the most comprehensible and easy to understand language. It provides significant information about this discipline to help develop a good understanding of it. This textbook will provide comprehensive knowledge to the readers.

The chapters below are organized to facilitate a comprehensive understanding of the subject:

Chapter - What is Horticulture?

The agriculture of plants for materials and food as well as for decoration is referred to as horticulture. Some of the important areas of horticulture are floriculture, pomology, viticulture, urban horticulture and arboriculture. This chapter has been carefully written to provide a brief introduction to the varied facets of horticulture.

Chapter - Diverse Aspects of Horticulture

A few of the fundamental concepts of horticulture are plug, bedding, bokashi and bolting. It also includes plant development, plant growth and plant embryogenesis. This chapter closely examines these fundamental concepts of horticulture to provide an extensive understanding of the subject.

Chapter - Horticultural Practices and Techniques

There are a number of techniques which are used in horticulture. They include plant propagation, propagation, tissue culture, grafting and plant breeding. The diverse applications of these practices and techniques of horticulture have been thoroughly discussed in this chapter.

Chapter - Organic Horticulture and Gardening

The practice of growing vegetables, fruits, flowers and ornamental plants using the essential principles of organic agriculture is referred to as organic horticulture. Some of the major sub-domains of organic horticulture are organic gardening, organic production of horticultural crops and organic fruit production. The topics elaborated in this chapter will help in gaining a better perspective about organic horticulture.

Chapter - Gardening

The practice of cultivating and growing plants as part of horticulture is known as gardening. Some of the major types of gardening are landscape gardening, container gardening, bonsai cultivation and sustainable gardening. This chapter has been carefully written to provide an easy understanding of these types of gardening.

Chapter - Disease Management of Horticultural Crops

Horticulture crops are susceptible to diverse range of diseases. The most common of them are bacterial speck, clubroot, anthracnose disease, floral malformation, leaf malformation, etc. This chapter closely examines these diseases as well as their management to provide an extensive understanding of the subject.

Beatrice Pollard

What is Horticulture? 1

- **Floriculture**
- **Pomology**
- **Viticulture**
- **Urban Horticulture**
- **Arboriculture**

The agriculture of plants for materials and food as well as for decoration is referred to as horticulture. Some of the important areas of horticulture are floriculture, pomology, viticulture, urban horticulture and arboriculture. This chapter has been carefully written to provide a brief introduction to the varied facets of horticulture.

Horticulture is the branch of plant agriculture dealing with garden crops, generally fruits, vegetables, and ornamental plants. As a general term, it covers all forms of garden management, but in ordinary use it refers to intensive commercial production. In terms of scale, horticulture falls between domestic gardening and field agriculture, though all forms of cultivation naturally have close links.

Horticulture is divided into the cultivation of plants for food (pomology and olericulture) and plants for ornament (floriculture and landscape horticulture). Pomology deals with fruit and nut crops. Olericulture deals with herbaceous plants for the kitchen, including, for example, carrots (edible root), asparagus (edible stem), lettuce (edible leaf), cauliflower (edible flower), tomatoes (edible fruit), and peas (edible seed). Floriculture deals with the production of flowers and ornamental plants; generally, cut flowers, pot plants, and greenery. Landscape horticulture is a broad category that includes plants for the landscape, including lawn turf, but particularly nursery crops such as shrubs, trees, and climbers.

The specialization of the horticulturist and the success of the crop are influenced by many factors. Among these are climate, terrain, and other regional variations.

Horticultural Regions

Temperate Zones

Temperate zones for horticulture cannot be defined exactly by lines of latitude or longitude but are usually regarded as including those areas where frost in winter occurs, even though rarely. Thus most parts of Europe, North America, and northern Asia are included, though some parts of the United States, such as southern California and Florida, are considered subtropical. A few parts of the north coast of the Mediterranean and the Mediterranean islands are also subtropical. In the Southern Hemisphere, practically all of New Zealand, a few parts of Australia, and the southern part of South America have temperate climates. For horticultural purposes altitude is also a factor; the lower slopes of great mountain ranges, such as the Himalayas and the Andes, are included. Thus the temperate zones are very wide and the range of plants that can be grown in them is enormous, probably greater than in either the subtropical or tropical zones. In the temperate zones are the great coniferous and deciduous forests: pine, spruce, fir, most of the cypresses, the deciduous oaks (but excluding many of the evergreen ones), ash, birch, and linden (lime).

The temperate zones are also the areas of the grasses—the finest lawns particularly are in the regions of moderate or high rainfall—and of the great cereal crops. Rice is excluded as being tropical, but wheat, barley, corn (maize), and rye grow well in the temperate zones.

Plants in the temperate zones benefit from a winter resting season, which clearly differentiates them from tropical plants, which tend to grow continuously. Bulbs, annuals, herbaceous perennials, and deciduous trees become more frost-resistant with the fall of sap and therefore have a better chance of passing the resting season undamaged. Another influence is the varying length of darkness and light throughout the year, so that many plants, such as chrysanthemums, have a strong photoperiodism. The chrysanthemum flowers only in short daylight periods, although artificial lighting in nurseries can produce flowers the year round.

Most of the great gardens of the world have been developed in temperate zones. Particular features such as rose gardens, herbaceous borders, annual borders, woodland gardens, and rock gardens are also those of temperate-zone gardens. Nearly all depend for their success on the winter resting period.

Tropical Zones

There is no sharp line of demarcation between the tropics and the subtropics. Just as many tropical plants can be cultivated in the subtropics, so also many subtropical and even temperate plants can be grown satisfactorily in the tropics. Elevation is a determining factor. For example, the scarlet runner bean, a common plant in temperate regions, grows, flowers, and develops pods normally on the high slopes of Mt. Meru in Africa near the Equator; it will not, however, set pods in Hong Kong, a subtropical situation a little south of the Tropic of Cancer but at a low elevation.

In addition to elevation, another determinant is the annual distribution of rainfall. Plants that grow and flower in the monsoon areas, as in India, will not succeed where the climate is uniformly wet, as in Bougainville in the Solomon Islands. Another factor is the length of day, the number of hours the Sun is above the horizon; some plants flower only if the day is long, but others make their growth during the long days and flower when the day is short. Certain strains of the cosmos plant are so sensitive to light that, where the day is always about 12 hours, as near the Equator, they flower when only a few inches high; if grown near the Tropics of Cancer or Capricorn, they attain a height of several feet, if the seeds are sown in the spring, before flowering in the short days of autumn and winter. Poinsettia is a short-day plant that may be seen in flower in Singapore on any day of the year, while in Trinidad it is a blaze of glory only in late December.

In the tropics of Asia and parts of Central and South America the dominant features of the gardens are flowering trees, shrubs, and climbers. Herbaceous plants are relatively few, but many kinds of orchids can be grown.

Vegetable crops vary in kind and quality with the presence or absence of periodic dry seasons. In the uniformly wet tropics, the choice is limited to a few root crops and still fewer greens. Sweet potatoes grow and bear good crops where the average monthly rainfall, throughout the year, exceeds 10 inches (25 centimetres); they grow even better where there is a dry season. The same can be said of taro, yams, and cassava. Tropical greens from the Malay Peninsula are not as good as those grown in South China, the Hawaiian Islands, and Puerto Rico. They include several spinaches, of which Chinese spinach or amaranth is the best; several cabbages; Chinese onions and chives; and several gourds, cucumbers, and, where there is a dry season, watermelons. Brinjals, or eggplants, peppers, and okra are widely cultivated. Many kinds of beans can be grown successfully, including the French bean from the American subtropics, the many varieties of the African cowpea, and yard-long bean. The yam bean, a native of tropical America, is grown for its edible tuber. In the drier areas the pigeon pea, the soybean, the peanut (groundnut), and the Tientsin green bean are important crops. Miscellaneous crops include watercress, ginger, lotus, and bamboo.

Propagation

Propagation, the controlled perpetuation of plants, is the most basic of horticultural practices. Its two objectives are to achieve an increase in numbers and to preserve the essential characteristics of the plant. Propagation can be achieved sexually by seed or asexually by utilizing specialized vegetative structures of the plant (tubers and corms) or by employing such techniques as cutting, layering, grafting, and tissue culture.

Seed Propagation

The most common method of propagation for self-pollinated plants is by seed. In self-pollinated plants, the sperm nuclei in pollen produced by a flower fertilize egg cells of a flower on the same plant. Propagation by seed is also used widely for many cross-pollinated plants (those whose pollen is carried from one plant to another). Seed is usually the least expensive and often the only means of propagation and offers a convenient way to store plants over long periods of time. Seed kept dry and cool normally maintains its viability from harvest to the next planting season. Some can be stored for years under suitable conditions. Seed propagation also makes it possible to start plants free of most diseases. This is especially true with respect to virus diseases, because it is almost impossible to free plants of virus infections and because most virus diseases are not transmitted by seed. There are two disadvantages to seed propagation. First, genetic variation occurs in seed from cross-pollinated plants because they are heterozygous. This means that the plant grown from seed may not exactly duplicate the characteristics of its parents and may possess undesirable characteristics. Second, some plants take a long time to grow from seed to maturity. Potatoes, for example, do not breed true from seed and do not produce large tubers the first year. These disadvantages are overcome by vegetative propagation.

The practice of saving seed to plant the following year has developed into a specialized part of horticulture. Seed technology involves all of the steps necessary to ensure production of seed with high viability, freedom from disease, purity, and trueness to type. These processes may include specialized growing and harvesting techniques, cleaning, and distribution. Relatively little tree and shrub seed is grown commercially; it is generally harvested from natural stands. Rootstock seed for fruit trees is often obtained as a by-product in fruit-processing industries. Seed growing and plant improvement are related activities. Thus many seed-producing firms actively engage in plant-breeding programs to accomplish genetic improvement of their material.

Harvesting of dry seed is accomplished by threshing. Seed from fleshy fruits is recovered through fermentation of the macerated (softened by soaking) pulp or directly from screening. Machines have been developed to separate and clean seed, based on size, specific gravity, and surface characteristics. Extended storage of seed requires low humidity and cool temperature.

Trade in seed requires quality control. For example, U.S. government seed laws require detailed labeling showing germination percentage, mechanical purity, amount of seed, origin, and moisture content. Seed testing is thus an important part of the seed industry. While most vegetable seed germinates readily upon exposure to normally favourable environmental conditions, many seed plants that are vegetatively (asexually) propagated fail to germinate readily because of physical or physiologically imposed dormancy. Physical dormancy is due to structural limitations to germination such as hard impervious seed coats. Under natural conditions weathering for a number of years weakens the seed coat. Certain seeds, such as the sweet pea, have a tough husk that can be artificially worn or weakened to render the seed coat permeable to gases and water by a process known as scarification. This is accomplished by a number of methods including abrasive action, soaking in hot water, or acid treatment. Physiologically imposed dormancy involves the presence of germination inhibitors. Germination in such seed may be accomplished by treatment to remove these inhibitors. This may involve cold stratification, storing seed at high relative humidity and low temperatures, usually slightly above freezing. Cold stratification is a prerequisite to the uniform germination of many temperate-zone species such as apple, pear, and redbud.

Vegetative Propagation

Asexual or vegetative reproduction is based on the ability of plants to regenerate tissues and parts. In many plants vegetative propagation is a completely natural process; in others it is an artificial one. Vegetative propagation has many advantages. These include the unchanged perpetuation of naturally cross-pollinated or heterozygous plants and the possibility of propagating seedless progeny. This means that a superior plant may be reproduced endlessly without variation. In addition, vegetative propagation may be easier and faster than seed propagation, because seed dormancy problems are eliminated and the juvenile nonflowering stage of some seed-propagated plants is eliminated or reduced.

Vegetative propagation is accomplished by use of (1) apomictic seed; (2) specialized vegetative structures such as runners, bulbs, corms, rhizomes, offshoots, tubers, stems, and roots; (3) layers and cuttings; (4) grafting and budding; and (5) tissue culture.

Apomixis

Apomixis, the development of asexual seed (seed not formed via the normal sexual process), is a form of vegetative propagation for some horticultural plants including Kentucky bluegrass, mango, and citrus. Virus-free progeny can be produced in oranges from a seed that is formed from the nucellus, a maternal tissue.

Vegetative Structures

Many plants produce specialized vegetative structures that can be used in propagation. These may be storage organs such as tubers that enable the plant to survive adverse conditions or organs adapted for natural propagation—runners or rhizomes—so that the plant may rapidly spread.

Bulbs consist of a short stem base with one or more buds protected by fleshy leaves. They are found in such plants as the onion, daffodil, and hyacinth. Bulbs commonly grow at ground level, though bulblike structures (bulbils) may form on aerial stems in some lilies or in association with flower parts, as in the onion. Buds in the axils (angle between leaf and stem) of the fleshy leaves may form miniature bulbs (bulblets) that when grown to full size are known as offsets. Corms are short, fleshy, underground stems without fleshy leaves. The gladiolus and crocus are propagated by corms. They may produce new cormels from fleshy buds. Rhizomes are horizontal, underground stems that are compressed, as in the iris, or slender, as in turf grasses. Runners are specialized aerial stems, a natural agent of increase and spread for such plants as the strawberry, strawberry geranium, and bugleweed (Ajuga). Tubers are fleshy enlarged portions of underground stem. The edible portion of the potato, the tuber, is also used as a means of propagation.

A number of plants form lateral shoots from the stem, which when rooted serve to propagate the plant. These are known collectively as offshoots but are often called offsets, crown divisions, ratoons, or slips.

Roots may also be structurally modified as propagative and food-storage organs. These tuberous roots, fleshy swollen structures, readily form shoots (called adventitious, because they do not form from nodes). The sweet potato and dahlia are propagated by tuberous roots. Shoots that rise adventitiously from roots are called suckers. The red raspberry is propagated by suckers.

Layering and Cutting

Propagation can be accomplished by methods in which plants are induced to regenerate missing parts, usually adventitious roots or shoots. When the regenerated part is still attached to the plant the process is called layerage, or layering; when the regenerating portion is detached from the plant the process is called cuttage, or cutting.

Layering often occurs naturally. Drooping black raspberry stems tend to root in contact with the soil. The croton, a tropical plant, is commonly propagated by wrapping moist sphagnum enclosed in plastic around a stem cut to induce rooting. After rooting, the stem is detached and planted. Though simple and effective, layering is not normally adapted to large-scale nursery practices.

Cutting is one of the most important methods of propagation. Many plant parts can be used; thus cuttings are classified as root, stem, or leaf. Stem cuttings are the most common.

The ability of stems to regenerate missing parts is variable; consequently plants may be easy or difficult to root. The physiological ability of cuttings to form roots is due to an interaction of many factors. These include transportable substances in the plant itself: plant hormones (such as auxin), carbohydrates, nitrogenous substances, vitamins, and substances not yet identified. Environmental factors such as light, temperature, humidity, and oxygen are important, as are age, position, and type of stem.

Although easy-to-root plants such as willow or coleus can be propagated merely by plunging a stem in water or moist sand, the propagation of difficult-to-root species is a highly technical process. To achieve success with difficult-to-root plants special care is taken to control the environment and encourage rooting. A number of growth regulators stimulate rooting. A high degree of success has been achieved with indolebutyric acid, a synthetic auxin that is applied to the cut surface. A number of materials known as rooting cofactors have been found that interact with auxin to further stimulate rooting, and these are sold as a hormone rooting compound.

Humidity control is particularly important to prevent death of the stem from desiccation before rooting is complete. The use of an intermittent-mist system in propagation beds has proved to be an important means of improving success in propagation by cuttings. These operate by applying water to the plant for a few seconds each minute.

Grafting

Grafting involves the joining together of plant parts by means of tissue regeneration. The part of the combination that provides the root is called the stock; the added piece is called the scion. When more than two parts are involved, the middle piece is called the interstock. When the scion consists of a single bud, the process is called budding. Grafting and budding are the most widely used of the vegetative propagation methods.

Stock cambium and scion cambium respond to being cut by forming masses of cells (callus tissues) that grow over the injured surfaces of the wounds. The union resulting from interlocking of the callus tissues is the basis of graftage. In dicots (e.g., most trees) cambium—a layer of actively dividing cells between xylem (wood) and phloem (bast) tissues—is usually arranged in a continuous ring; in woody members, new layers of tissue are produced annually. Monocot stems (e.g., lilacs, orchids) do not possess a continuous cambium layer or increase in thickness; grafting is seldom possible.

The basic technique in grafting consists of placing cambial tissues of stock and scion in intimate association, so that the resulting callus tissue produced from stock and scion interlocks to form a living continuous connection. A snug fit can be obtained through the tension of the split stock and scion or both. Tape, rubber, and nails can be used to achieve close contact. In general, grafts are only compatible between the same or closely related species. Success in grafting depends on skill in achieving a snug fit. Warm temperatures (80°–85° F [27°–30° C]) increase callus formation and improve "take" in grafting. Thus grafts using dormant material are often stored in a warm, moist place to stimulate callus formation.

In grafting and budding, the rootstock can be grown from seed or propagated asexually. Within a year a small amount of scion material from one plant can produce hundreds of plants.

Grafting has uses in addition to propagation. The interaction of rootstocks may affect the performance of the stock through dwarfing or invigoration and in some cases may affect quality. Further, the use of more than one component can affect the disease resistance and hardiness of the combination.

Grafting as a means of growth control is used extensively with fruit trees and ornamentals such as roses and junipers. Fruit trees are normally composed of a scion grafted onto a rootstock. Sometimes an interstock is included between the scion and stock. The rootstock may be grown from seed (seedling rootstock) or asexually propagated (clonal rootstock). In the apple, a great many clonal rootstocks are available to give a complete range of dwarfing; rootstocks are also available to invigorate growth of the scion cultivar.

Tissue-culture techniques utilizing embryos, shoot tips, and callus can be used as a method of propagation. The procedure requires aseptic techniques and special media to supply inorganic elements; sugar; vitamins; and, depending on the tissue, growth regulators and organic complexes such as coconut milk, yeast, and amino-acid extract.

Embryo culture has been used to produce plants from embryos that would not normally develop within the fruit. This occurs in early-ripening peaches and in some hybridization between species. Embryo culture can also be used to circumvent seed dormancy.

A shoot tip, when excised and cultured, may produce roots at the base. This technique is employed for the purpose of producing plants free of disease. Certain orchids are rapidly multiplied by this method. Cultured shoot tips form an embryo-like stage that can be sectioned indefinitely to build up large stocks rapidly. These bulblike bodies left unsectioned develop into small plantlets. A similar procedure is used with the carnation, in which the shoot tip forms a cell mass that can be subdivided.

Callus-tissue culture—a very specialized technique that involves growth of the callus, followed by procedures to induce organ differentiation—has been successful with a number of plants including carrot, asparagus, and tobacco. Used extensively in research, callus culture has not been considered a practical method of propagation. Callus culture produces genetic variability because in some cases cells double their chromosome number. In rice and tobacco, mature plants have been obtained from callus formed from pollen. These plants have half the normal number of chromosomes.

Breeding

The isolation and production of superior types known as cultivars are the very keystones of horticulture. Plant breeding, the systematic improvement of plants through the application of genetic principles, has placed improvement of horticultural plants on a scientific basis. The raw material of improvement is found in the great variation that exists between cultivated plants and related wild species. The incorporation of these changes into cultivars adapted to specific geographical areas requires a knowledge of the theoretical basis of heredity and art and the skill to discover, perpetuate, and combine these small but fundamental differences in plant material.

The goal of the plant breeder is to create superior crop varieties. The cultivated variety, or cultivar, can be defined as a group of crop plants having similar but distinguishable characteristics. The term cultivar has various meanings, however, depending on the mode of reproduction of the crop. With reference to asexually propagated crops, the term cultivar means any particular clone considered of sufficient value to be graced with a name. With reference to sexually propagated crops, the concept of cultivar depends on the method of pollination. The cultivar in self-pollinated crops is basically a particular homozygous genotype, a pure line. In cross-pollinated crops the cultivar is not necessarily typified by any one plant but sometimes by a particular plant population, which at any one time is composed of genetically distinguishable individuals.

Environmental Control

Control of the natural environment is a major part of all forms of cultivation, whatever its scale. The basic processes involved in this task have already been described in a preceding section on the principles of gardening, and these also apply to horticulture. The scale, intensiveness, and economic risk in commercial gardening and nurseries, however, often require approaches markedly different from those of the small home garden; and some of these are described here.

The intensive cultivation practiced in horticulture relies on extensive control of the environment for all phases of plant life. The most basic environmental control is achieved by location and site: sunny or shady sites, proximity to bodies of water, altitude, and latitude.

Structures

Various structures are used for temperature control. Cold frames, used to start plants before the normal growing season, are low enclosed beds covered with a removable sash of glass or plastic. Radiant energy passes through the transparent top and warms the soil directly. Heat, however, as long-wave radiation, is prevented from leaving the glass or plastic cover at night. Thus heat that builds up in the cold frame during the day aids in warming the soil, which releases its heat gradually at night to warm the plants. When supplemental heat is provided, the structures are called hotbeds. At first, supplemental heat was supplied by respiration through the decomposition of manure or other organic matter. Today, heat is provided by electric cables, steam, or hot-water pipes buried in the soil.

Greenhouses are large hotbeds, and in most cases the source of heat is steam. While they were formerly made of glass, plastic films are now extensively used. Modern greenhouse ranges usually have automatic temperature control. Summer temperatures can be regulated by shading or evaporative "fan-and-pad" cooling devices. Air-conditioning units are usually too expensive except for scientific work. Greenhouses with precise environmental controls are known as phytotrons. Other environmental factors are controlled through automatic watering, regulation of light and shade, addition of carbon dioxide, and the regulation of fertility.

Shade houses are usually walk-in structures with shading provided by lath or screening. Summer propagation is often located in shade houses to reduce excessive water loss by transpiration.

Temperature Control

A number of temperature-control techniques are used in the field, including application of hot caps, cloches, plastic tunnels, and mulches of various types. Hot caps are cones of translucent paper or plastic that are placed over the tops of plants in the spring. These act as miniature greenhouses. In the past small glass sash called cloches were placed over rows to help keep them warm. Polyethylene tunnels supported by wire hoops that span the plants are now used for the same purpose. As spring advances the tunnels are slashed to prevent excessive heat buildup. In some cases the plastic tunnels are constructed so that they can be opened and closed when necessary. This technique is widely used in Israel for early production of vegetables.

Mulching is important in horticulture. Whether in the form of a topdressing of manure or compost or plastic sheeting, mulches offer the grower the various benefits of economical plant feeding, conservation of moisture, and control of weeds and erosion. Winter mulches are commonly used to protect such sensitive and valuable plants as strawberries and roses.

The storage of perishable plant products is accomplished largely through the regulation of their temperature to retard respiration and microbial activity. Excess water loss can be prevented by controlling humidity. Facilities that utilize the temperature of the atmosphere are called common storage. The most primitive types take advantage of the reduced temperature fluctuations of the soil by using caves or unheated cellars. Aboveground structures must be insulated and ventilated. Complete temperature-regulated storages utilizing refrigeration and heating are now common for storage of horticultural products. The regulation of oxygen and carbon dioxide levels along with the regulation of temperature is known as controlled-atmosphere storage. Rooms are sealed so that gaseous exchange can be effectively controlled. Many horticultural products, such as fruit, can be kept fresh for as long as a year under these controlled conditions.

Frost Control

Frost is one of the high-risk elements for commercial growers, and the problem is accentuated by the fact that growers are striving to produce early-season crops. The precautions are consequently far more elaborate and costly than those of the domestic garden. Frost is especially damaging to perennial fruit crops in the spring—because flower parts are sensitive to freezing injury—and to tender transplants. The two weather conditions that produce freezing temperatures are rapid radiational cooling at night and introduction of a cold air mass with temperatures below freezing. Radiation frost occurs when the weather is clear and calm; air-mass freezes occur when it is overcast and windy.

Frost-control methods involve either reduction of radiational heat loss or conservation or addition of heat. Radiational heat loss may be reduced by hot caps, cold frames, or mulches. Heat may also be added from the air. Wind machines that stir up the air, for example, provide heat when temperature inversions trap cold air under a layer of warm air. These have been used extensively in citrus groves. Heat may be added directly by using heaters, usually fueled with oil. Sprinkler irrigation can also be used for frost control. The formation of ice is accompanied by the release of large amounts of heat, which maintains plants at the freezing temperature as long as the water is being frozen. Thus continuous sprinkling during frosty nights has been used to protect strawberries from frost injury.

Frost injury to transplants can be prevented through processes that increase the plant's ability to survive the impact of unfavourable environmental stress. This is known as hardening off. Hardening off of plants prior to transplanting can be accomplished by withholding water and fertilizer, especially nitrogen. This prevents formation of succulent tissue that is very frost-tender. Gradual exposure to cold is also effective for hardening. Induced cold resistance in crops such as cabbage, for example, can have a considerable effect; unhardened cabbages begin to show injury at 28 °F (−2.2 °C), while hardened plants withstand temperatures as low as 22 °F (−5.6 °C).

Light Control

Light has a tremendous effect on plant growth. It provides energy for photosynthesis, the process by which plants, with the aid of the pigment chlorophyll, synthesize carbon compounds from water and carbon dioxide. Light also influences a great number of physiological reactions in plants. At energy values lower than those required for photosynthesis, light affects such processes as dormancy, flowering, tuberization, and seed-stalk development. In many cases these processes are affected by the length of day; the recurrent cycle of light is known as the photoperiod.

The control of light in horticultural practices involves increasing energy values for photosynthesis and controlling day length. Light is controlled in part by site and location. In the tropics day length approaches 12 hours throughout the year, whereas in polar regions it varies from zero to 24 hours. Light is also partly controlled by plant distribution and density.

Supplemental illumination in greenhouses increases photosynthesis. The cost of power to supply the artificial light, however, makes this impractical for all but crops of the highest value. Fluorescent lights are the most efficient for photosynthesis; special lights, rich in the wavelengths required, are now available.

Extension of day length through supplemental illumination and shading is common practice in the production of greenhouse flower crops, which are often induced to flower out of season. Artificial lengthening of short days, or interruption of the dark period, promotes flowering in long-day plants such as lettuce and spinach and prevents flowering of short-day plants such as chrysanthemums. Similarly, during naturally long days, shading to reduce day length prevents flowering of long-day plants and promotes flowering of short-day plants. The manipulation of day length is standard practice to control flowering of greenhouse chrysanthemums throughout the year. Tungsten lights have proved very effective for extending day length because they are rich in the red end of the spectrum that affects the photoperiodic reaction. Extending the day length is a relatively affordable practice because only a low light intensity is required. The same effects can be obtained through interruption of the dark period, even with light flashes. Decreasing day length is usually accomplished by simply covering the plants with black shade cloth.

Soil Management

The principles involved here are again similar to those of home gardening. But the financial considerations of horticulture naturally require a more scientific approach to soil care. To be successful, the grower must ensure the economic use of every square yard of ground, especially because the cost of sound horticultural land is among the highest of any in agriculture. Crop rotation is planned to ensure that the soil is not depleted of essential chemicals by repeated use of one type of plant in the same plot. Soil analysis is employed so that any such depletion can be rectified promptly. Fertilizers are applied in a precise routine and, of course, in a variety beyond the reach or needs of the ordinary gardener. They are frequently applied through leaves or stems in the form of chemical sprays.

Water Management

Depending on the terrain, water management may involve extensive works for irrigation and drainage. While the home gardener may well be content with a rough-and-ready appraisal of the wetness or dryness of the soil, horticulture is more exacting. Production of the high-quality fruits and vegetables demanded by the modern market requires a precise all-year balance of soil moisture, adjusted to the needs of the particular crop. These considerations apply whether the grower is situated in a high-rainfall area of Europe or in the parched land of the southwestern United States or Israel.

There are a number of general methods of land irrigation. In surface irrigation water is distributed over the surface of soil. Sprinkler irrigation is application of water under pressure as simulated rain. Subirrigation is the distribution of water to soil below the surface; it provides moisture to crops by upward capillary action. Trickle irrigation involves the slow release of water to each plant through small plastic tubes. This technique is adapted both to field and to greenhouse conditions.

Removal of excess water from soils can be achieved by surface or subsurface drainage. Surface drainage refers to the removal of surface water by development of the slope of the land utilizing systems of drains to carry away the surplus water. In subsurface drainage open ditches and tile fields intercept groundwater and carry it off. The water enters the tiling through the joints, and drainage is achieved by gravity feed through the tiles.

Pest Control

Horticultural plants are subject to a wide variety of injuries caused by other organisms. Plant pests include viruses, bacteria, fungi, higher plants, nematodes, insects, mites, birds, and rodents. Various methods are used to control them. The most successful treatments are preventive rather than curative.

Control of pests is achieved through practices that prevent harm to the plant and methods that affect the plant's ability to resist or tolerate intrusion by the pathogen. These can be classified as cultural, physical, chemical, or biological.

Traditional practices that reduce effective pest population include the elimination of diseased or infected plants or seeds (roguing), cutting out of infected plant parts (surgery), removal of plant debris that may harbour pests (sanitation), and alternating crops unacceptable to pests (rotation). Any of a number of techniques can be employed to render the environment unfavourable to the pest, such as draining or flooding and changing the soil's level of acidity or alkalinity.

Physical methods can be used to protect the plant against intrusion or to eliminate the pest entirely. Physical barriers range from the traditional garden fence to bags that protect each fruit, a common practice in Japan. Heat treatment is used to destroy some seed-borne pathogens and is a standard soil treatment in greenhouses to eliminate soil pests such as fungi, nematodes, and weed seed. Cultivation and tillage are standard practices for weed control.

The horticultural industry is now dependent upon chemical control of pests through pesticides, materials toxic to the pest in some stage of its life cycle. Commercial growers of practically all horticultural crops rely on complete schedules utilizing many different compounds. Pesticides are usually classed according to the organism they control: for example, bactericide, fungicide, nematicide, miticide, insecticide, rodenticide, and herbicide.

Selectivity of pesticides, the ability to discriminate between pests, is a relative concept. Some non-selective pesticides kill indiscriminately; most are selective to some degree. Most fungicides, for example, are not bactericidal. The development of highly selective herbicides makes it possible to destroy weeds from crops selectively. Selectivity can be achieved through control of dosage, timing, and method of application.

Plant pests can also be controlled through the manipulation of biological factors. This may be achieved through directing the natural competition between organisms or by incorporating natural resistance to the whole plant. The introduction of natural parasites or predators has been a successful method for the control of certain insects and weeds. Incorporation of genetic resistance is an ideal method of control. Thus breeding for disease and insect resistance is one of the chief goals of plant breeding programs. A major obstacle to this method of control is the ability of pathogens (disease-producing organisms) to mutate easily and attack previously resistant plants.

Growth Regulation by Chemicals

Control of plant growth through growth-regulating materials is a modern development in horticulture. These materials have resulted from basic investigations into growth and development, as well as systematic screening of materials to find those that affect differentiation and growth. This field was given great impetus by the discovery of a class of plant hormones known as auxins, which affect cell elongation.

Auxins have been correlated with inhibition and stimulation of growth as well as differentiation of organs and tissues. Such processes as cell enlargement, leaf and organ separation, budding, flowering, and fruit set (the formation of the fruit after pollination) and growth are influenced by auxins. In addition, auxins have been associated with the movement of plants in response to light and gravity. Auxin materials are used in horticulture for the promotion of rooting, fruit setting, fruit thinning, and fruit-drop control.

Gibberellins are a group of related, naturally occurring compounds of which only one, gibberellic acid, is commercially available. Gibberellins have many effects on plant development. The most startling is the stimulation of growth in many compact or dwarf plants. Minute applications transform bush to pole beans or dwarf to normal corn. Perhaps the most widespread horticultural use has been in grape production. The application of gibberellin is now a regular practice for the culture of the 'Thompson seedless' cultivar ("Sultanina") of grapes to increase berry size. In Japan applications of gibberellic acid are used to induce seedlessness in certain grapes.

Cytokinins are a group of chemical substances that have a decisive influence on the stimulation of cell division. In tissue culture high auxin and low cytokinin give rise to root development; low auxin and high cytokinin encourage shoot development.

Ethylene, a hydrocarbon compound, acts as a plant hormone to stimulate fruit ripening as well as rooting and flowering of some plants. An ethylene-releasing compound, 2-chloroethylphosphonic acid, has many horticultural applications, of which the most promising may be uniform ripening of tomatoes and the stimulation of latex flow in rubber.

Many compounds that inhibit growth hormones have application in horticulture. For example, a number of materials that inhibit formation of gibberellins by the plant cause dwarfing. These include chlorinated derivatives of quaternary ammonium and phosphonium compounds. Many of these have applications in floriculture. Growth retardants such as succinic acid–2,2-dimethyl-hydrazide, a gibberellin suppressor, have applications in horticulture from a wide array of effects that include dwarfing and fruit maturity. The growth inhibitor maleic hydrazide has been effective in preventing the sprouting of onions and potatoes.

Ornamental Horticulture

Ornamental horticulture consists of floriculture and landscape horticulture. Each is concerned with growing and marketing plants and with the associated activities of flower arrangement and landscape design. The turf industry is also considered a part of ornamental horticulture. Although flowering bulbs and flower seed represent an important component of agricultural production for the Low Countries of Europe, ornamentals are relatively insignificant in world trade.

Floriculture has long been an important part of horticulture, especially in Europe and Japan, and accounts for about half of the nonfood horticultural industry in the United States. Because flowers and pot plants are largely produced in plant-growing structures in temperate climates, floriculture is largely thought of as a greenhouse industry; there is, however, considerable outdoor culture of many flowers.

The industry is usually very specialized with respect to its crop; the grower must provide precise environmental control. Exact scheduling is imperative since most floral crops are seasonal in demand. Because the product is perishable, transportation to market must function smoothly to avoid losses.

The floriculture industry involves the grower, who mass-produces flowers for the wholesale market, and the retail florist, who markets to the public. The grower is often a family farm, but, as in all modern agriculture, the size of the growing unit is increasing. There is a movement away from urban areas, with their high taxes and labour costs, to locations with lower tax rates and a rural labour pool and also toward more favourable climatic regions (milder temperature and more sunlight). The development of airfreight has emphasized interregional and international competition. Flowers can be shipped long distances by air and arrive in fresh condition to compete with locally grown products.

The industry of landscape horticulture is divided into growing, maintenance, and design. Growing of plants for landscape is called the nursery business, although a nursery refers broadly to the growing and establishment of any young plant before permanent planting. The nursery industry involves production and distribution of woody and herbaceous plants and is often expanded to include ornamental bulb crops—corms, tubers, rhizomes, and swollen roots as well as true bulbs. Production of cuttings to be grown in greenhouses or for indoor use (foliage plants), as well as the production of bedding plants, is usually considered part of floriculture, but this distinction is fading. While most nursery crops are ornamental, the nursery business also includes fruit plants and certain perennial vegetables used in home gardens, for example, asparagus and rhubarb.

Next to ornamental trees and shrubs, the most important nursery crops are fruit plants, followed by bulb crops. The most important single plant grown for outdoor cultivation is the rose. The type of nursery plants grown depends on location; in general (in the Northern Hemisphere) the northern areas provide deciduous and coniferous evergreens, whereas the southern nurseries provide tender broad-leaved evergreens.

The nursery industry includes wholesale, retail, and mail-order operations. The typical wholesale nursery specializes in relatively few crops and supplies only retail nurseries or florists. The wholesale nursery deals largely in plant propagation, selling young seedlings and rooted cuttings, known as "lining out" stock, of woody material to the retail nursery. The retail nursery then cares for the plants until growth is complete. Many nurseries also execute the design of the planting in addition to furnishing the plants.

Bulb Crops

The bulb crops include plants such as the tulip, hyacinth, narcissus, iris, daylily, and dahlia. Included also are nonhardy bulbs used as potted plants indoors and summer outdoor plantings such as amaryllises, anemones, various tuberous begonias, caladiums, cannas, dahlias, freesias, gladioli, tigerflowers, and others. Hardy bulbs, those that will survive when left in the soil over winter, include various crocuses, snowdrops, lilies, daffodils, and tulips.

Many bulb crops are of ancient Old World origin, introduced into horticulture long ago and subjected to selection and crossing through the years to yield many modern cultivars. One of the most popular is the tulip. Tulips are widely grown in gardens as botanical species but are especially prized in select forms of the garden tulip (which arose from crosses between thousands of cultivars representing several species). Garden tulips are roughly grouped as early tulips, breeder's tulips, cottage tulips, Darwin tulips, lily-flowered tulips, triumph tulips, Mendel tulips, parrot tulips, and others. The garden tulips seem to have been developed first in Turkey but were spread throughout Europe and were adopted enthusiastically by the Dutch. The Netherlands has been the centre of tulip breeding ever since the 18th century, when interest in the tulip was so intense that single bulbs of a select type were sometimes valued at thousands of dollars. The collapse of the "tulipmania" left economic scars for decades. The Netherlands remains today the chief source of tulip bulbs planted in Europe and in North America. The Netherlands has also specialized in the production of related bulbs in the lily family and provides hyacinth, narcissus, crocus, and others. The Dutch finance extensive promotion of their bulbs to support their market. Years of meticulous growing are required to yield a commercial tulip bulb from seed. Thorough soil preparation, high fertility, constant weeding, and careful record keeping are part of the intensive production, which requires much hand labour. Bulbs sent to market meet specifications as to size and quality, which assure at least one year's bloom even if the bulb is supplied nothing more than warmth and moisture. The inflorescence (flowering) is already initiated and the necessary food stored in the bulb. Under less favourable maintenance than prevails in the Netherlands, a subsequent year's bloom may be smaller and less reliable; it is not surprising therefore that tulip-bulb merchants suggest discarding bulbs after one year and replanting with new bulbs to achieve maximum yield.

Herbaceous Perennials

Garden perennials include a number of herbaceous species grown for their flowers or occasionally used as vegetative ground covers. Under favourable growing conditions the plants persist and increase year after year. The biggest drawback to perennials as compared with annuals is that they must be maintained throughout the growing season but have only a limited flowering period. Typical perennials are hollyhocks, columbines, bellflowers, chrysanthemums, delphiniums, pinks, coralbells, phlox, poppies, primroses, and speedwells.

Perennials are often produced and sold as a sideline to other nursery activities; some are sold through seed houses. Perennial production could be undertaken on a massive scale, with attendant economies, but the market is neither large enough nor predictable enough (except for the greenhouse growing of such cut flowers as chrysanthemums and carnations) to interest most growers.

Shrubs

Production of ornamental shrubs is the backbone of the nursery trade in Europe and the United States. The nursery business is about equally divided between the production of (1) coniferous evergreens such as yew, juniper, spruce, and pine; (2) broad-leaved evergreens such as rhododendron, camellia, holly, and boxwood; (3) deciduous plants such as forsythia, viburnum, berberis, privet, lilac, and clematis; and (4) roses.

Fields of specialization have evolved within the ornamental shrub industry. Some firms confine activity mostly to production of "lining out" stock, which must be tended several years before reaching salable size.

The field grower may, in turn, specialize in mass growing for the wholesale trade only. The field plantings are tended until they attain marketable size. Because of the time required to produce a marketable crop and because of rising labour costs, this phase of the nursery industry involves economic hazards. But wholesale growing escapes the high overhead of retail marketing in urban areas, and, although many growers do sell stock at the nursery, they generally avoid the expensive merchandising required of the typical urban-area garden centre. Growers are especially interested in laboursaving technology and are turning to herbicidal control of weeds and shortcut methods for transplanting.

There is a well-established trade in container-grown stock—that is, nursery stock grown in the container in which it is sold. This practice avoids transplanting and allows year-round sales of plant material.

Roses

The production of roses is probably the most specialized of all shrub growing; the grower often deals solely in rose plants. Most are bud-grafted onto rootstocks (typically Rosa multiflora). This is the only way to achieve rapid and economical increase of a new selection to meet market demands. Large-scale production of roses has tended to centre in areas where long growing seasons make rapid production possible.

Because the budding operation calls for skilled hand labour and because field maintenance is expensive, few economies can be practiced in the production of roses. But distribution techniques that do offer certain economies have been developed. These include covering the roses with coated paper or plastic bags instead of damp moss to retain humidity and applying a wax coating to stems of dormant stock to inhibit desiccation.

Trees

Ornamental shade trees are usually grown and marketed in conjunction with shrubs. The 20th-century migration of people in many countries to suburban areas, coupled with the construction of houses on cleared land, has made shade trees an increasingly important part of the nursery trade. As interest in shade and ornamental trees increased, creation of improved cultivars followed. There is still some activity in transplanting native trees from the woodlot, and some are still grown from genetically unselected seed or cuttings; but more and more, like roses and shrubs before them, trees are vegetatively propagated as named cultivars, and many are patented.

The design and planning of landscapes has become a distinct profession that in many cases is only incidentally horticultural. Landscape architecture in its broadest sense is concerned with all aspects of land use. As a horticulturist, the landscape architect uses plants along with other landscape materials—stone, mortar, wood—as elements of landscape design. Unlike the materials of the painter or sculptor, plants are not static but change seasonally and with time. The colour, form, texture, and line of plants are used as design elements in the landscape. Plant materials are also manipulated as functional materials to control erosion, as surface materials, and for enclosures to provide protection from sunlight and wind.

Landscape architecture originated in the design of great estates, and home landscape is still an integral part of landscape architecture. More recently, however, landscape architecture has begun to include larger developments such as urban and town planning, parks both formal and "wild," public buildings, industrial landscaping, and highway and roadside development.

Floriculture

Floriculture or flower farming is the study of growing and marketing flowers and foliage plants. Floriculture includes cultivation of flowering and ornamental plants for direct sale or for use as raw materials in cosmetic and perfume industry and in the pharmaceutical sector. It also includes production of planting materials through seeds, cuttings, budding and grafting. In simpler terms floriculture can be defined as the art and knowledge of growing flowers to perfection. The persons associated with this field are called floriculturists.

Worldwide more than 140 countries are involved in commercial Floriculture. The leading flower producing country in the world is Netherlands and Germany is the biggest importer of flowers. Countries involved in the import of flowers are Netherlands, Germany, France, Italy and Japan while those involved in export are Colombia, Israel, Spain and Kenya. USA and Japan continue to be the highest consumers.

Floriculture in India

Floriculture is an age old farming activity in India having immense potential for generating gainful self-employment among small and marginal farmers. In the recent years it has emerged as a profitable agri-business in India and worldwide as improved standards of living and growing consciousness among the citizens across the globe to live in environment friendly atmosphere has led to an increase in the demand of floriculture products in the developed as well as in the developing countries worldwide. The production and trade of floriculture has increased consistently over the last 10 years. In India, Floriculture industry comprises flower trade, production of nursery plants and potted plants, seed and bulb production, micro propagation and extraction of essential oils. Though the annual domestic demand for the flowers is growing at a rate of over 25% and international demand at around Rs 90,000 crore India's share in international market of flowers is negligible. However, India is having a better scope in the future as there is a shift in trend towards tropical flowers and this can be gainfully exploited by country like India with high amount of diversity in indigenous flora.

After liberalization the Government of India identified floriculture as a sunrise industry and accorded it 100 percent export oriented status. The liberalization of industrial and trade policies paved the way for the development of export oriented production of cut flowers. The new seed policy has already made it feasible to import planting material of international varieties. Floriculture products mainly consist of cut flowers, pot plants, cut foliage, seeds bulbs, tubers, rooted cuttings and dried flowers or leaves. The important floricultural crops in the international cut flower trade are rose, carnation, chrysanthemum, gerbera, gladiolus, orchids, anthurium, tulip and lilies.

According to statistics indicated in the Handbook on Horticulture Statistics 2014, the total area under flower crops in 2012-13 was 232.70 thousand hectares. Total area under floriculture in India is second largest in the world and only next to China. Production of flowers was estimated to be 1729.2 MT of loose flowers and 76731.9 million (numbers) of cut flowers in 2012-13. Fresh and Dried cut flowers dominate floriculture exports from India.

Among states, Karnataka is the leader in floriculture with about 29,700 hectares under floriculture cultivation. Other major flower growing states are Tamil Nadu and Andhra Pradesh in the South, West Bengal in the East, Maharashtra in the West and Rajasthan, Delhi and Haryana in the North.

The expert committee set up by Govt. of India for promotion of export oriented floriculture units has identified Bangalore, Pune, New Delhi and Hyderabad as the major areas suitable for such activity especially for cut flowers. Of the four zones identified as potential centers for flower production namely Bangalore, Hyderabad, Pune and New Delhi, the area around Bangalore and Pune have got the advantage of ideal climatic conditions where the temperature ranges between 15 to 30°C. In view of this, the units established in these locations do not require either cooling or heating system. As a result maximum number of units has been established in these locations. There are more than 300 export oriented units in India. APEDA (Agricultural and Processed Food Products Export Development Authority) is the registering authority for such units.

Marketing

In India Marketing of cut flowers is much unorganized. In most of the Indian cities flowers are brought to wholesale markets, which mostly operate in open yards. From here the flowers are distributed to the local retail outlets which more often than not operate in the open on-road sides, with different flowers arranged in large buckets. In the metropolitan cities, however, there are some good florist show rooms, where flowers are kept under controlled temperature conditions, with considerable attention to value added service.The government is now investing in setting up of auction platforms, as well as organized florist shops with better storage facilities to prolong shelf life. The packaging and transportation of flowers from the farms to the retail markets at present is very unscientific. The flowers, depending on the kind, are packed in gunny bags, bamboo baskets, simple cartons or just wrapped in old newspapers and transported to markets by road, rail or by air. However, the government has provided some assistance for buying refrigerated cargos and built up a large number of export oriented units with excellent facilities of pre-cooling chambers, cold stores and reefer vans.

According to a study titled, 'Indian Floriculture Industry: The Way Ahead' released by the apex industry body ASSOCHAM, India's floriculture industry is growing at a compounded annual growth rate of about 30%, and is likely to cross Rs 8,000 crore mark by 2015. Currently, the floriculture industry in India is poised at about Rs 3,700 crore with a share of a meagre 0.61% in the global floriculture industry which is likely to reach 0.89% by 2015.

Export Constraints

In spite of an abundant and varied production base, India's export of floricultural product is not encouraging. The low performance is attributed to many constraints like non-availability of air space in major airlines. The Indian floriculture industry is facing with a number of challenges mainly related to trade environment, infrastructure and marketing issues such as high import tariff, low availability of perishable carriers, higher freight rates and inadequate refrigerated and transport facilities. At the production level the industry is faced with challenges mostly related to availability of basic inputs including quality seeds and planting materials, efficient irrigation system and skilled manpower. In order to overcome these problems, steps must be taken to reduce import duty on planting material and equipment, reduce airfreight to a reasonable level, provide sufficient cargo space in major airlines and to establish model nurseries for supplying genuine planting material. Training centres should be established for training the personnel in floriculture and allied areas. Exporters should plan and monitor effective quality control measures right from production to post harvesting,storage, and transportation.

Government Programmes and Policies

Department of Agriculture and Cooperation under the Ministry of Agriculture is the nodal organization responsible for development of the floriculture sector. It is responsible for formulation and implementation of national policies and programmes aimed at achieving rapid agricultural growth through optimum utilization of land, water, soil and plant resources of the country. Production of cut flowers for exports is also a thrust area for support. The Agricultural and Processed Food Products Export Development Authority (APEDA), the nodal organization for promotion of agri exports including flowers, has introduced several schemes for promoting floriculture exports from the country. These relate to development of infrastructure, packaging, market development, subsidy on airfreight for export of cut flowers and tissue-cultured plants, database up-gradation etc. The 100% Export Oriented Units are also given benefits like duty free imports of capital goods. Import duties have also been reduced on cut flowers, flower seeds, tissue-cultured plants, etc. Setting up of walk in type cold storage has been allowed at the International airports for storage of export produce.

Initiatives have also been launched for the benefit of exporters by providing cold storage and cargo handling facility for perishable products at various international airports. Direct subsidy up to 50 percent is also available in cold storage units. Besides, subsidy is also provided by APEDA on improved packaging materials to promote their use. To attract entrepreneurship in floriculture sector, NABARD is providing financial assistance to hi-tech units at reasonable interest rates.

Several schemes have been initiated by the Government for promotion and development of the floriculture sector including "Integrated Development of Commercial Floriculture" which aims at improvement in production and productivity of traditional as well as cut flowers through availability of quality planting material, production of off season and quality flowers through protected cultivation, improvement in post harvest handling of flowers and training persons for a scientific floriculture. Many state governments have set up separate departments for promotion of floriculture in their respective states.

Research work on floriculture is being carried out at several research institutions under the Indian Council of Agricultural Research and Council of Scientific and Industrial Research, in the horticulture/floriculture departments of State Agricultural Universities and under the All India Coordinated Floriculture Improvement Project with a network of about twenty (20) centres. The key focus areas are crop improvement, standardization of agro-techniques including improved propagation methods, plant protection and post harvest management. In recent years, however, technologies for protected cultivation and tissue culture for mass propagation have also received attention. A large number of promising varieties of cut flowers have been developed. All these efforts indicate the government's commitment for improving the sector and creating a positive environment for entrepreneurship development in the field.

Demand and Supply

The demand for flowers is seasonal as it is in most countries. The demand for flowers has two components: a steady component and a seasonal component. The factors which influence the demand are to some extent different for traditional and modern flowers.

- Traditional Flowers

The steady demand for traditional flowers comes from the use of flowers for religious purposes, decoration of homes and for making garlands and wreaths. This demand is particularly strong in Kerala, Karnataka, Tamil Nadu, Odisha and West Bengal, as the use of flowers for above mentioned purposes is part of their local culture. The bulk of seasonal demand comes from festivals and marriages. The demand is generally for specific flowers.

- Modern Flowers

The bulk of the steady demand for modern flowers comes from institutions like hotels, guest houses and marriage gardens. The demand is concentrated in urban areas. With increasing modernization and globalization the demand for modern flowers from the individual consumers is likely to grow enormously as the trend of "say it with flowers" is increasing and the occasions which call for flower giving will continue to present themselves. Although there is an increasing demand for modern flowers from individuals, institutions continue to be the dominant buyers in the market. The price of these flowers also depends on their demand and varies accordingly.

Green House Technology for Flower Production

In present scenario of increasing demand for cut flowers protected cultivation in green houses is the best alternative for using land and other resources more efficiently. In protected environment suitable environmental conditions for optimum plant growth are provided which ultimately provide quality products. Green House is made up of glass or plastic film, which allows the solar radiations to pass through but traps the thermal radiations emitted by plants inside and thereby provide favourable climatic conditions for plant growth. It is also used for controlling temperature, humidity and light intensity inside. On the basis of basic material used, building cost and technology used, green houses can be of three types:

- Low-cost greenhouse: The low-cost green house is made of polythene sheet of 700 gauge supported on bamboos with twines and nails. Its size depends on the purpose of its utilization and availability of space. The temperature within greenhouse increases by 6-10oC more than outside.
- Medium-cost greenhouse: With a slightly higher cost greenhouse can be framed with GI pipe of 15 mm bore. This greenhouse has a covering of UV -stabilized polythene of 800 gauge. The exhaust fans are used for ventilation which are thermostatically controlled. Cooling pad is used for humidifying the air entering the chamber. The greenhouse frame and glazing material have a life span of about 20 years and 2 years respectively.
- Hi-tech greenhouse: In this type of green house the temperature, humidity and light are automatically controlled according to specific plant needs. These are indicated through sensor or signal-receiver. Sensor measures the variables, compare the measurement to a standard value and finally recommend to run the corresponding device. Temperature control system consists of temperature sensor heating/cooling mechanism and thermostat operated fan. Similarly, relative humidity is sensed through optical tagging devices. Boiler operation, irrigation and misting systems are operated under pressure sensing system. This modern structure is highly expensive, requiring qualified operators, maintenance, care and precautions. However, these provide best conditions for export quality cut flowers and are presently used by large number of export units.

Challenges Faced by Floriculture Producers

Growers face many challenges including:

- Declining margins: While prices have remained steady over the past several years, most input costs have risen steadily. To remain profitable, growers have had to become more efficient in production and management.
- Environment: Environmental issues are a major concern for growers. Growers have responded by re-using irrigation water, reducing pesticide and fertilizer use and reducing greenhouse runoff.
- Pest control: Concerns over pesticide use by the public and producers alike, along with pesticide resistance and the loss of approved pesticides, have prompted growers to adopt alternative pest control methods. Integrated pest management (IPM) is playing a larger

role in greenhouse pest control. Many growers are now using biological or bio-rational control methods to supplement or replace existing pesticides.

- Employment: Labour is an important element in production. Bedding plant and cut flower growers face labour costs of up to one third of gross sales. Although increased mechanization is a necessary element of global competition, the industry continues to be a major agricultural employer.
- Urban-rural conflicts: Urban-rural conflicts are a fact of life for most agriculture in the Province. Some municipalities look upon floriculture as more of a factory production industry rather than agriculture. Most municipalities have zoning regulations concerning the maximum site coverage for greenhouses.
- Capital costs: Modern, state-of-the-art greenhouse operations can cost up to $200 per square metre. This represents a barrier to entry for many potential growers. Field-grown cut flowers and bedding plant production have much lower capital costs, so they are often entry level crops.
- Seasonal demand: The demand for fresh floriculture products is seasonal and the product is very perishable. Large numbers of people want to buy flowers for special occasions or holidays like St. Valentine's Day, Easter, Mother's Day and Christmas. Growers must time their production to meet these periods of high demand. Some growers have 30% of their annual sales in a three week period in spring.

Pomology

Pomology or Fruit farming is the growing of fruit crops, including nuts, primarily for use as human food.

The subject of fruit and nut production deals with intensive culture of perennial plants, the fruits of which have economic significance (a nut is a fruit, botanically). It is one part of the broad subject of horticulture, which also encompasses vegetable growing and production of ornamentals and flowers. This topic places further arbitrary limitations in that it does not encompass a number of very important perennial fruit crops covered elsewhere, including vanilla, coffee, and the oil-producing tung tree and oil.

Botanists define a fruit in broad terms as the fleshy or dry ripened ovary surrounding the seed of a plant. A pomologist, or specialist in the science and practice of fruit growing, defines it somewhat more narrowly as the fleshy edible part of a perennial plant associated with development of the flower. A nut is any seed or fruit consisting of a kernel, usually oily, surrounded by a hard or brittle shell. Most edible nuts—e.g., almond, walnut, cashew, pecan, pistachio, etc.—are well known as dessert nuts. Not all nuts are edible. Some, used as sources of oil or fat, may be regarded as oil seeds; others are used for ornament. The botanical definition of a nut, based on features of form and structure (morphology), is more restrictive: a hard, dry, one-celled, one-seeded fruit that does not split open at maturity. Among the nuts that fit both the botanical and popular conception are the acorn, chestnut, and filbert; other so-called nuts may be botanically a seed (Brazil nut), a legume (peanut [groundnut]), or a drupe (almond and coconut).

Improvements in technology and consolidation of the fruit and nut industries in the most favoured climates of the world have been responsible for a steady increase in yield. Thus, the total acreage or number of plants devoted to various fruit and nut crops has dropped, remained about the same, or not risen in proportion to the increase in the respective crop production.

Although fruit- and nut-growing enterprises cover great ranges of climates and plant materials, their technologies have many common problems and practices. The most significant of these are discussed below.

The Variety: Its Propagation and Improvement

The first step in establishing a fruit- or nut-growing industry is the selection of individual plants with high productivity and a superior product. Such an individual is a horticultural variety. If it is multiplied vegetatively from rooted cuttings, from root pieces that throw shoots, or by graftage, each plant in the group (called a clone) that results is identical with the others. Nearly all commercially important perennial fruit and nut crops are clonally propagated; i.e., their varieties are multiplied vegetatively by one means or another. Some nut crops, such as the wild pecan, cashew, black walnut, hickory, and chestnut still come from trees that grow at random from seed; hence, character and quality tend to vary.

Many important varieties of fruit plants were selected generations ago. The Sultanina (Thompson Seedless) grape, the Lob Injir (Calimyrna) fig, and the Gros Michel banana have obscure origins; planted by the millions since selection, each specimen is actually a vegetative continuation of the selected individual growing on an independent root system. But regardless of the age of a fruit-growing industry, or the perfection of some of the selected varieties, a continuing search for new varieties is essential. There is always room for improvement in climatic adaptability, in insect and disease resistance, and in the solution of special horticultural or marketing problems. In fact, government experiment stations over the world now stress scientific breeding for improvement of market quality and yield of key fruit and nut crops.

Not only are varietal selection and improvement a continuing need but so also is the maintenance of existing varieties. Although an improved vegetative mutation of a variety is exceptional, the opportunities for accidental multiplication of degenerate (low-quality) mutants increase in proportion to the number of specimens of the variety. As a result, care is taken to propagate a clone only from superior individuals, and in the case of citrus, where mutation is especially common, further precautions are necessary. There are, of course, occasional mutations that may greatly improve a variety and these are sought, selected, and propagated.

Vegetative propagation technique varies with the individual fruit plant. Date, banana, and pineapple are multiplied by use of offshoots or suckers. Grape, fig, olive, currant, and blueberry are usually propagated from cuttings. Strawberry and black raspberry reproduce vegetatively by special organs—the former by stolons or runners, the latter by cane tip rooting or layering. Many kinds of fruit trees must be grafted or budded on especially grown rootstocks because the species to be multiplied does not root itself easily; apple, pear, peach, mango, and citrus are examples of this group. Many nut trees have a single taproot with but few branching roots, necessitating a deep hole and special care in transplanting.

Today's trend is toward a smaller tree in most fruit crops, particularly the apple and pear, and toward closer planting in hedgerow style, with carefully regulated fertilization and irrigation. This increases production per acre, lowers labour cost, increases early yields, and facilitates access in maintenance and harvesting. This approach, in fact, has been used for decades in Europe. Labour is the largest element of cost in fruit and nut production. Every means is exploited to reduce, facilitate, or eliminate hand labour.

With most fruit species a period of one to two years intervenes between the time a cutting is rooted and the time the plant is ready for setting in the field, or between graftage or budding and field planting. During this interval the plants remain in a nursery where they can be given intensive culture in rows. Pineapple and banana planting materials, however, do not require nursery care before field planting.

In choosing fruit varieties, the grower must (1) recognize the relative adaptabilities of available varieties to the climatic and soil conditions of his farm and (2) select a group that satisfies both his management needs and the market demands from those best adapted to his conditions. For instance, an apple producer in the northeastern U.S. may raise four varieties: Milton, McIntosh Red, Red Delicious, and Rome Beauty. The main harvest seasons for these succeed each other at two-week intervals; this helps him extend the harvest period and make efficient use of his labour. The first two varieties cross-fertilize satisfactorily, as do the last two. The first of these varieties is usually marketed without storage, while the storage seasons of the others are of increasing length. This helps the grower to extend his marketing period.

Cultivation

Site Selection

The site of a fruit-growing enterprise is as significant in determining its success as the varieties grown. In fact, variety and site together set a ceiling on the productivity and profit that can be realized under the best management. In most developed fruit regions microclimatic conditions (climate at plant height, as influenced by slight differences in soil, soil covering, and elevation) and soil conditions are the two components of a site that determine its desirability for a fruit-growing enterprise. Sometimes (particularly with highly perishable fruits) transportation to market must also be considered.

Local conditions at a site that expose it to unusual frost hazard are as detrimental to citrus in Florida as they are to peach trees in New Zealand and apple trees in the south of England. In regions and sites where temperatures during the season may drop no more than a few degrees below freezing, artificial frost protection is sometimes used. This is accomplished by open-flame burning (petroleum bricks, logs, etc.) or heating of metal objects with oil, gas, propane, electricity, etc. (stones or stacks that radiate heat). Another technique is the spraying of water on plants (e.g., strawberries) as long as the temperature is below freezing.

For highest productivity, most fruit trees must root extensively to a depth of three feet (one metre) or more. Heavy subsoil or other conditions causing imperfect internal drainage may result in shallow, weak root systems that do not take water and nutrients efficiently from the soil. In semi-arid and arid regions, accumulation of saline soils in a subsurface layer sometimes limits rooting of fruit trees, causes abnormal foliar symptoms, and reduces yields. Tiling and surface ditching help decrease water accumulation in poorly drained subsoils and reduce wet spots in otherwise satisfactory sites. Special control of irrigation procedures and periodic leaching may alleviate the worst salt effects in saline soils. Choice of tolerant species, varieties, and rootstocks may make fruit growing economical on imperfectly drained or mildly saline sites, though plants rarely perform as well as they do on sites free from these difficulties. Coconuts, however, tolerate saline soil conditions near tropical saltwater coasts.

Once selected, a site is cleared, levelled (if needed), and cultivated. Then drainage, irrigation, and road systems are installed as required. In rolling or sloping terrain, where contour planting is needed to control erosion and conserve moisture, the locations of the plant or row positions are determined by the contour terraces and waterways established. In old lands, nematode or other pest populations make fumigation necessary before planting. In some problem California soils, giant plows and treaded tractors turn the soil to depths of three to six feet (one to two metres). In very infertile sites, or sites where the physical condition of the surface soil is poor, it may be helpful to grow a succession of leguminous cover crops for a year or more before planting and/or apply a fertilizer containing major fertilizer elements (nitrogen, potassium, phosphorus, calcium, sulfur) and all or certain trace elements (iron, manganese, boron, zinc, copper, molybdenum) and lime, based on a soil test.

Planting and Spacing Systems

Growth, flowering habits, and light requirements on the one hand, and management problems on the other, determine the most satisfactory planting plan for a fruit- and nut-growing enterprise. There is a trend toward use of dwarfing stocks, growth control chemicals, or closer planting and training, or all of them to get the highest yields and best operation efficiency possible on a unit of ground.

Low-growing crops such as strawberry and pineapple are usually managed in beds containing several rows, or in less formal matted rows. In an acre of strawberries, 200,000 or more plants may occupy the matted rows. A pineapple plantation with two-row beds, having plants one foot (0.3 metre) apart in rows two feet (0.6 metre) apart totals 15,000 to 18,000 plants per acre (37,000 to 44,000 per hectare). With such dense populations, intense competition for light, water, and nutrients causes smaller average fruit size. Nevertheless, the total yield per unit of land is usually greater than it would be with lower plant numbers.

The spacing of grapevines along a trellis row and of trees planted in hedgerows involves the same group of problems. Maximum vineyard production frequently results with vine distances of eight to nine feet (2.4 to 2.7 metres; 600 ± per acre [1,500 per hectare]). The trend for peach trees and spur-type apple strains is hedgerows 14 feet (4.2 metres) apart or closer, in rows 18 to 20 feet (5.4 to 6 metres) apart.

With those species and varieties that require cross-pollination by insects, the planting plan must take those special needs into account. This is a problem with apple, pear, plum, and sweet cherry orchards. At least two varieties that cross-fertilize successfully must be planted in association with each other.

Training and Pruning

Pruning is the removal of parts of a plant to influence growth and fruitfulness. It is an important fruit-growing practice. Primary attention is given to form in the first few years after fruit trees or vines are planted. Form influences strength and longevity of the mature plant as well as efficiency of other fruit-growing practices; pruning for form is called training. As the plant approaches maximum fruitfulness and fills its allotted space, maintenance pruning for various purposes becomes increasingly important.

The grape may be trained following one of two systems: (1) spur system, cutting growth of the previous season (canes) to short spurs, (2) long-cane system, permitting canes to remain relatively long. Whether a spur or long-cane system is followed depends on the flowering habit of the variety. Relatively small trees that respond favourably to severe annual pruning (e.g., the peach and Kadota fig) are usually trained to create an open-centred tree with a scaffold of four or five main branches that originate on a short trunk and branch a number of times to provide fruiting wood. Annual renewal pruning can be reasonably efficient under these circumstances. Larger trees that do not respond favourably to heavy annual pruning are trained best to a system that encourages the main leader branch to grow erect to a height of eight to 10 feet (2.4–3.0 metres), with four or five main lateral branches at intervals on its sides forming the scaffold that carries fruiting wood up and out; this is called a modified leader system. The central leader type of tree, with one main leader up through the centre and many side branches, is common for pear and apple planted in hedgerows, and possibly for other fruits and nuts as the close-planted hedgerow system is more widely adopted.

The principal reasons for maintenance pruning are: (1) to permit efficient spraying and harvesting operations, (2) to maintain satisfactory light exposure for most of the leaves, and (3) to create a satisfactory balance between flowering and leaf surface.

To reduce hand labour costs, larger commercial fruit growers use machine pruning on many types of fruits. Peach, apple, pear, and other fruits usually planted in hedgerows are mowed across the top and sides by machine, then thinned out as needed by a follow-up crew using pneumatic clippers and hand-powered saws, operating from hydraulically manipulated scaffolds or lifts of various types.

Soil Management, Irrigation and Fertilization

Soil Management

Two soil management practices (1) clean cultivation and chemical weed control or both and (2) permanent sod culture, illustrate contrasting purposes and effects. In clean cultivation or chemical weed control, the surface soil is stirred periodically throughout the year or a herbicide is used to kill vegetation that competes for nutrients, water, and light. Stirring increases the decomposition rate of soil organic matter and thereby releases nitrogen and other nutrients for use by the fruit crop. It may also provide some improvement in water penetration. On the other hand, laying bare the soil surface exposes it to erosion; destruction of organic matter eventually lowers fertility and causes soil structure to change from loose and friable to tight and compacted. Though sod culture minimizes the destructive processes and may permit a modest increase in fertility, the sod itself competes with fruit plants for water and nutrients and may even compete for light. As a result, permanent sod culture is practical only with tree crops that are normally rather low in vegetation, such as apple, pear, sweet cherry, nuts, and mango. Competition from established sod may be detrimental to vigorously growing fruit plants like grape, peach, and raspberry unless adequate fertilizer and water are supplied.

Because each of these soil management systems has advantages and disadvantages, modifying or complementary practices are often used; for example, cover cropping, mulching, and chemical control of vegetation with or without strip sod in the row middles. In fact, the trend is toward mowed sod middles with strip chemical control under the trees and with overhead sprinklers during hot dry weather. Sprinklers not only provide water but tend to cool the plants and give fruit of better market quality without aggravating diseases. Cultivation combined with winter cover cropping has been used widely in grape, peach, cherry, bush fruit, and citrus plantings, as well as with other species. Mulching is the addition of undecomposed plant materials such as straw, hay, or processors' refuse to the soil under the plants. In orchards, mulching materials are most often applied under trees maintained in permanent sod. Strip in-row chemical control of vegetation in commercial fruit plantings has almost taken over as an economical and sound practice.

Irrigation

In semi-arid and arid regions, irrigation is necessary. Probably the maximum demand occurs in date gardens, because they expose a large leaf surface the year around under conditions of high evaporation and practically no rainfall. Irrigation in humid climates is generally being provided increasingly during extended dry periods that occur at one time or another during most growing seasons. For example, large acreages of banana are irrigated on coastal lowlands of the torrid tropics where annual rainfall exceeds 60 inches (1,500 millimetres).

Fertilization

Needs of perennial fruit plants for fertilizers depend on the natural fertility of the soil supporting them and on their individual requirements. Of the essential elements, supplemental nitrogen is almost always needed; potassium supplements may be needed, even in some desert areas. Although strawberry, grape, peach, and a few other fruits have responded favourably to phosphorus, and although its application has been recommended, the phosphorus requirement of woody plants is low and deficiency is rather rare. Calcium deficiency may be more common than realized; lime is often desirable to reduce soil acidity and because of other indirect benefits. Inadequate magnesium in the soil has been noted by workers studying a wide range of fruit species. Of the trace elements, zinc, iron, and boron are most likely to be deficient, but copper, manganese, and molybdenum deficiencies also are being reported for some fruits in some regions. Iron deficiency is difficult to control in orchards where soils have high alkalinity. Granulated fertilizers in modern close-planted commercial orchards are usually broadcast by machine a month or two before growth starts. Additional nitrogen sometimes is applied in heavy crop years to apple, pear, and citrus.

Crop Enhancement

Pollination

The stimulus of pollination, fertilization, and seed formation is needed to get good size, shape, and flavour of most of the fruits. (Banana, pineapple, and some citrus and fig varieties are exceptions). Transfer of pollen from the anthers (male) to the stigmas (female) is accomplished in nature either by insects or by movement in air. It is common practice to bring beehives into the orchard during bloom. Rainy cold weather during bloom with little or no sunshine can deter activity of the honey bee (the key insect pollinator) and reduce fruit set appreciably. This is one of the main problems not fully solved by fruit researchers. Hand-pollination by daubing collected and preserved pollen onto the stigma (as is done with date palms) sometimes is practiced for other fruits, but this approach is not widespread.

Thinning

Removal of flowers or young fruit (thinning) is done to permit the remaining fruits to grow more rapidly and to prevent development of such a large crop that the plant is unable to flower and set a commercial crop the following year. Thinning is done by hand, mechanically, or chemically. With the date, the pistillate flower cluster is reduced in size at the time of hand-pollination. In the case of certain table grape varieties, some clusters are cut off. With the Thompson seedless grape, a combination of girdling the trunk bark and judicious application of gibberellin (growth regulating) sprays at blossoming gives excellent full bunches.

Young peach fruits are thinned by striking the branches with a padded pole or by shaking the entire tree for a few seconds with a well-padded motor-driven shaker arm grasping the trunk. Hand thinning of young apple and peach fruits once was also a common practice, but because of the expense and difficulty, there has been increasing use of chemical sprays as a substitute. Two kinds of sprays are used: (1) mildly caustic sprays applied during bloom, such as Elgetol in arid regions, or (2) sprays of growth-regulating substances such as 3-CPA (2,3-chlorophenoxy propionamide) applied within a few weeks after bloom in areas with late frosts.

Pest Control and Preservation

In many fruit enterprises, pest control is the most expensive and time-consuming growing practice. Where the concentration of fruit farms in an area warrants it, individual efforts are complemented by legislative measures including quarantine regulations to force removal of pest-laden, unattended orchards. Sometimes the most economical control procedure is biological in nature. There is increased research today to find and multiply parasites that kill fruit crop pests. Such biological methods are necessary as political pressures increase for banning DDT and other chemicals. Selection of varieties that are immune, resistant to attack, or tolerant to specific pests, is a biological control procedure also widely used. Chemical control procedures, however, are relied on most heavily. Air-blast spray or mist-application machinery covering 70 acres (28 hectares) of trees or more in a day is now in common use.

Harvesting and Packing

The proper time to remove a fruit from the tree or plant varies with each fruit and is governed by whether the product will be sold and consumed within hours, or stored for weeks, months, or even a year. Most fruits are harvested as close as possible to the time they are eaten. A few, of which banana and pear are outstanding examples, may be harvested while immature and still ripen satisfactorily. Orange, grapefruit, and some varieties of avocado may be "stored" on the tree for several months after they have attained good quality; this method cuts costs in handling and marketing.

Many fruits, including apple, pear, orange, lemon, and grapefruit, may drop from the tree during the last part of the maturation period. Preharvest drop of these fruits can be delayed by application of dilute sprays of growth-regulating substances like naphthaleneacetic acid (NAA). The chemical spray Alar [N-(dimethylamino) succinamic acid] applied four to six weeks after bloom on apple not only reduces fruit drop at harvest but increases red colour, firmness, and return bloom the next year, in addition to other advantages.

For the fresh market, most tree and bush fruits are still harvested by hand. For processing, drying, and occasionally for fresh market, mechanical motor-driven tree and bush shakers with appropriate catching belts, bins, pallets, and electric lifts reduce harvesting and handling labour. In years to come, machinery may make it possible to machine-harvest most fruits, with no more, and possibly less, damage than with hand picking.

The public has become increasingly particular about the appearance and quality of the product it buys. Hence, store managers and suppliers seek the best grades of fruits and nuts available, and growers make every effort to produce crops with attractive colour and smooth finish. Fruits are packed by government-controlled grades such as Fancy or Extra Fancy within given size limits and are so labelled on the carton or box, together with the source. Most fruits and nuts not meeting this standard of quality are processed or sent through channels using the lower grades and off sizes.

Small packages of plastic foam or wood pulp base holding four to six fruits covered and heat sealed with polyethylene plastic film are popular. These are delivered to stores in corrugated cartons holding a few dozen packages. Citrus, apples, and whole nuts or kernels also are packaged in polyethylene bags and delivered in cartons. Loose fruit may be sold in cell cartons and tray packs consisting of stacked form-fitting pulp trays in a "bushel size" box. Every effort is made to eliminate bruising.

Large truck-pulled containers with individually motor-driven refrigeration units, with or without controlled atmosphere (CO_2-O_2, to retard ripening), are loaded at the fruit source and trucked to their destination or are loaded on ships by derrick for overseas shipment. These sealed containers are also being used increasingly for bananas to reduce labour and handling and to deliver the product in better condition.

Air shipment of "vine- and tree-ripe" fruit (strawberries, figs, sweet cherries, pineapples, avocados) to distances as far as from California to Europe in a day or less is becoming increasingly common with the much larger and faster cargo planes and reduced air-freight prices.

Postharvest Physiology of Fruits

Fruit ripening is a form of senescence and signifies the final stage in fruit development. A fleshy fruit is the enlarged ovary of a flower (avocado) or additional floral parts such as in apple, pear, and pineapple. Usually fertilization, and sometimes pollination alone, stimulate the floral parts causing a rapid cell division that leads to differentiation and the formation of the fruit structure. During this stage fruits consist of small, young cells filled with protoplasm. When the young fruit has been stimulated, presumably by plant hormones that originate from the embryonic seeds, rapid cell expansion takes place. During this stage fruits gain rapidly in size and weight. The cells develop small cavities or spaces in their tissue (become vacuolated) and begin the process of foodstuff accumulation, which lends fruits their compositional diversity. Banana, apple, and date, for example, accumulate mainly carbohydrates. Avocado and olive store fatty materials. Important constituents of most fruits are organic acids such as malic acid, found in apple and pear; citric acid, found in citrus fruits and pineapple; and tartaric acid, found in grapes. Fruits are usually low in protein.

After cell expansion has slowed and become nominal, fruits enter the stage of maturity and undergo preparation for ripening. Some crops, such as pear and avocado, are harvested at the so-called mature-green state and allowed to ripen afterward. Most fruits are at a stage of incipient ripening before they are picked. Ripening is marked by rapid and dramatic changes that give fruits their attractive and edible character. Some of the familiar changes are softening, which results from degradation of cell wall substances; disappearance of a green background, because of chlorophyll degradation (as in pear, apple, and banana); appearance of coloured pigments such as the carotenoids—orange-yellow—and anthocyanins—red (as in orange, mango, and strawberry); a decrease in acidity and increase in the sugar content (orange, apple); and emission of the volatile substances that give many fruits their distinct aroma (as in banana, pear, and apple). In climacteric fruits (e.g., banana, pear, apple), ripening is accompanied by increased respiration. In nonclimacteric fruit (e.g., strawberry, cherry) this phenomenon does not occur.

It is thought that the transition from the mature to the ripe stage is brought about by certain "ripening" enzymes. Protein molecules act as catalysts. The activity of these enzymes leads first to various ripening reactions, and then to gradual deterioration of the fruit tissue.

Because ripening leads to tissue breakdown, fruits are considered a highly perishable commodity. Different fruits have varying degrees of postharvest longevity. While strawberries last only a week to 10 days, for example, apples or lemons can be stored successfully for as long as several months.

Postharvest life of fruits can be extended by refrigeration with or without a modified oxygen–carbon dioxide atmosphere. Most temperate-zone fruits can be held safely at 32 to 41 °F (0 to 5 °C), but many subtropical and tropical fruits, including lemon, avocado, banana, and mango, show signs of injury from being chilled in prolonged cold storage and consequently fail to ripen properly.

Bananas do not tolerate temperatures below 53 °F (12 °C), while several avocado varieties can be stored at temperatures as low as 46 °F (8 °C).

Fruit life can be extended further by both refrigeration and controlled atmosphere (CA) storage in which oxygen is kept at about 5 percent and carbon dioxide at 1 to 3 percent, while temperature is held at a level best suited to the particular fruit. So-called CA storage is common today for apples and pears and is being adapted to other fruits. Controlled atmosphere and refrigeration in conjunction with the removal of ethylene gas (which emanates from fruits and speeds ripening) helps slow the ripening process considerably. Golden Delicious apples and some pears are shipped in polyethylene containers in which a desirable, modified atmosphere is created by the respiring fruit.

Drying is a standard practice for stabilizing the market movement of dates, figs, raisin grapes, prunes, and apricots. Canning is of paramount importance to the pineapple, peach, and pear industries (these fruits can be dried as well), and freezing is a means of stabilizing some of the most perishable fruits, including strawberry, raspberry, and blueberry.

Nuts are susceptible to mold, souring, staleness, discoloration, and rancidity. Cured and dried nuts are kept in prolonged cold storage under controlled temperature and humidity levels. Nuts also are stored and sold in vacuum packs of carbon dioxide-enriched atmosphere.

Waste Materials

Apple wood is excellent for fireplace use, and cherry and certain other fruit woods are used for the finest household furniture. The dried residue from processing apples and citrus is made into feed for conditioning livestock for market, as are waste materials from many processed fruits. Apple pomace (waste material) is spread on the orchard floor with a manure spreader to help in soil conditioning and as a source of minerals.

Nutshells have many uses. Filbert shells are made into plywood, artificial wood, and linoleum; a mixture of shells with powdered coal and lignite makes cinder blocks; shells are used in making poisonous gases and gas masks, and as fuel and mulch. Cashew shell liquid, a skin irritant, is made into resins for varnishes; kills mosquito larvae; can be impregnated in wood as a varnish to preserve against insect attack; is used in automotive brake linings and clutch facings; is used as a laminating agent for paper, cloth, and glass fibres; and is used to treat cement floors and synthetic rubber to retard deterioration. Finely ground black-walnut-shell flour is used in plastic molding powder; as a glue extender; to prevent overheating of drills; to "sand"-blast jet engines; for polishing, burnishing, and deburring metal parts; for cleaning foundry molds; and to spray on tires for better traction. Pecan shells are used in place of gravel in cement walks and driveways; as fuel; as mulch and as a soil conditioner; in livestock bedding; as filler for fertilizers, feeds, etc.; in the manufacture of tanning agents, with charcoal and abrasives in hand soap; as a filler in plastic and veneer wood; and many of the same uses as black walnut shells. Some nutshells are made into beads, marbles, buttons, carving tools, ink, and ornament. The India clearing nut is cut open and rubbed on the inside of earthenware that will contain drinking water; the juice coagulates the water impurities which sink to the bottom. The nuts of the betel palm in the Far East and of the kola tree in West Africa are chewed for their stimulatory effects.

Viticulture

Viticulture is the process of grape production. Grapes are grown around the globe apart from the Antarctica, and they have high adaptability properties to different environments. Grapes are fruits which are used to produce wine. The people who study the science behind grape production are called viticulturists. Viticulturists study pests and diseases to control, irrigation, fertilization, fruit development, and characteristics of grapes. They are also responsible in managing the canopy on which the grapes grow, when to prune, and when to harvest the fruit. Winemakers also liaise with viticulturists in wine production because the best vines give the best wine. Different varieties of grapes are now approved by the European Union as the real grapes for wine production, because of their characteristics. Some of the wines produced using grapes include the red wine, which is produced from gulp of black and red grapes. Grapes are also used in making raisins.

Uses of Grapes

Grapes are berries. The system through which grapes are organized is called clustering. Some clusters contain grapes which are compact berries while others are spread out. Long clustered grapes spread out while short clusters are packed together. Some grapes ripen together making harvesting easier while others ripen individually within the same cluster. Raisins are made from dried grapes which could be seedless and they are used in cooking, brewing, and baking. Fertilization occurs in one to four seeds within a berry. Inside each grape, there is a rachis which allows the grape to get nutrients and water. One plant can produce 100 to 200 grapes. The skin of grape weighs about 5% to 20% of the total weight of the grape. The skin when ripe contains tannin and some aromatic substances. When the grape is ripe, the tannin is used to formulate color and body shape.

Conditions for Growing Grapes

Each type of grape has a favorable climatic condition. Temperatures and rainfall are uncontrollable conditions and each year will see the growth of unique grapes. During summer times, the right temperatures help in the ripening of the grapes. These cool temperatures allow for the resting of the fruit. Very cold temperatures cause frosting of the fruit. The grapevine should also be on a hillside, and this allows for sunrays falling at an angle where all vines get sunshine, unlike on flat ground where the sunrays do not reach every plant well. Slopes also offer better drainage. Quality soil is important in determining the health and growth of the vine. The right kind of soil allows for the plant's roots to develop well. Soil with loose texture, moderate fertility, and great drainage are favorable for grape growth.

Problems Facing Grape Farming

Weather conditions like wind, frost, a lot of heat, and poor drainage are some of the factors that adversely affect the quality grape harvest. Diseases are also rife among the grape farmrs. Some of the viticulture hazards include oidium, downy mildew, fan leaf, plant virus, and phylloxera. Spraying the plants with copper sulfate treats the problem of downy mildew. Some species from North America have developed resistance to this disease. Fanleaf has no cure so affected are removed. Oidium can be terminal and does well in the shade and cooler temperatures.

Grape Harvesting

A green harvest is done on immature grapes to reduce yield. Removing those green grapes helps the remaining ones to put their energy into developing better and healthy grapes. A healthy, vigorous and more mature flavor compound is achieved through removing some bunches of green grapes. The practice of intermixing different varieties of grapes produces a wine called a field blend. Field blend allows for effortlessly blending of different varieties with different genetic makeup. In Germany, field blend is called a mixed set.

Precision Viticulture

Precision viticulture is precision farming applied to optimize vineyard performance, in particular maximizing grape yield and quality while minimizing environmental impacts and risk. This is accomplished by measuring local variation in factors that influence grape yield and quality (soil, topography, microclimate, vine health, etc.) and applying appropriate viticulture management practices (trellis design, pruning, fertilizer application, irrigation, timing of harvest, etc.). Precision viticulture is based on the premise that high in-field variability for factors that affect vine growth and grape ripening warrants intensive management customized according to local conditions. Precision viticulture depends on new and emerging technologies such as global positioning systems (GPS), meteorologic and other environmental sensors, satellite and airborne remote sensing, and geographic information systems (GIS) to assess and respond to variability.

Precision viticulture is unique in its emphasis on vineyard management according to local variation, and in its use of science and technology to accomplish this. While Australian viticulturists are generally recognized as leaders in precision viticulture, and while viticulturists worldwide have embraced the approach, the fundamental concepts have deep roots in the traditions of Old World winemaking regions. *Terroir*, a related French concept, refers to the special geographic qualities or "sense of place" embodied in the wine produced in a particular region.

Precision agriculture emphasizes "doing the right thing, in the right place, at the right time", and is practical for viticulture because of high local variability of conditions within vineyards, and because of responsiveness to intensive management in terms of increased grape yield and quality. According to CSIRO, Australia "Typically grape yield varies eight to ten-fold under uniform management"; "patterns of yield variation are stable in time and driven by soil and topographic variation"; and "patterns of variation in fruit quality tend to be similar to those for yield, suggesting opportunities for zonal management and selective harvest". Australian precision viticulture has focused on yield monitoring, whereas California precision viticulture has focused on remote sensing.

Technology for Characterizing Vineyard Variation

Precision viticulture uses a broad set of enabling technologies to observe and respond to vineyard variability:

- Global Positioning Systems (GPS) provide satellite-based georeferencing for mapping vineyard environmental variability.
- Meteorologic Stations monitor climatic factors important for vine growth and grape ripening, including temperature, precipitation, humidity, and wind.

- Remote Sensing from satellite and airborne platforms provides images depicting vineyard conditions, for example vine productivity from normalized difference vegetation index (NDVI).
- Digital Elevation Models (DEM) provide detailed topographic information.
- High Resolution Soil Surveys provide detailed information about soil fertility and hydrologic characteristics.
- Relational Databases organize environmental and economic information.
- Geographic Information Systems (GIS) provide digital tools for map-based analysis.
- Other Environmental Sensors monitor important biophysical factors such as solar radiation, soil moisture, and temperature regimes.

Management Practices

Precision viticulture draws upon a variety of management approaches, including zonal management, in which different areas of the vineyard are managed according to their unique conditions, and adaptive management, in which different management practices are applied according to observed needs and improved knowledge. Trellis design, in terms of row orientation and geometry of vine support, and pruning practices can be tailored to optimize vine health, to protect grapes from frost, sunburn, and mildew damage, and to ensure even grape ripening. Irrigation and fertilizer application schedules, pest management, and selective harvest based on timing of ripening can all be managed to minimize costs and maximize vineyard performance based on observed needs. Increasingly, precision viticulture, with its focus on management according to local variability, is coupled with organic farming, with its focus on environmentally friendly practices without the use of chemical pesticides and fertilizers, and with sustainable agriculture, with emphasis on long-term environmental stewardship and economic viability.

Ripeness in Viticulture

In viticulture, ripeness is the completion of the ripening process of wine grapes on the vine which signals the beginning of harvest. What exactly constitutes ripeness will vary depending on what style of wine is being produced (sparkling, still, fortified, rosé, dessert wine, etc.) and what the winemaker and viticulturist personally believe constitutes ripeness. Once the grapes are harvested, the physical and chemical components of the grape which will influence a wine's quality are essentially set so determining the optimal moment of ripeness may be considered the most crucial decision in winemaking.

There are several factors that contribute to the ripeness of the grape. As the grapes go through veraison, sugars in the grapes will continue to rise as acid levels fall. The balance between sugar (as well as the potential alcohol level) and acids is considered one of the most critical aspects of producing quality wine so both the must weight and "total acidity", as well as the pH of the grapes, are evaluated to determine ripeness. Towards the end of the 20th century, winemakers and viticulturists began focusing on the concept of achieving "physiological" ripeness in the grapes-described as a more complete ripeness of tannins and other phenolic compounds in the grapes that contribute to the color, flavor and aroma of wine.

What Happens to a Grape as it Ripens?

Pinot noir grapes in the early stages of veraison. As the grapes ripen, the concentration of phenolic compounds like anthocyanins replaces the green color of chlorophyll in the grape berries which makes them black instead.

If ripening is broadly defined as the development of wine grapes, then it could be said that ripening is happening throughout the continuous annual cycle of the grapevine. More narrowly defined, ripening begins at the inception of *veraison*. At this point (which is normally 40–60 days after fruit set though it may be longer in cooler climates), the grapes are hard and green with low sugar levels and very high levels of mostly malic acids. During *veraison*, which may last from 30–70 days depending on the climate and other factors, the grapes go through several changes which impact their sugar, acid, tannin and mineral composition. The concentration of phenolic compounds in the skin, most notably anthocyanins for red wine grapes, replace the green color of chlorophyll as the grape berries themselves change color.

The increase of sugars in the grapes comes from the storage of carbohydrates in the roots and trunk of the grapevines as well as through the process of photosynthesis. Sucrose produced by photosynthesis is transferred from the leaves to the berries as it is broken down into glucose and fructose molecules. The rate of this build up will depend on several factors including the climate (such as a string of cloudy weather which prohibits sunlight from reaching the vine) as well as the potential yield size of grape clusters and young vine shoot tips which compete for the resources of the mother grapevine. As the concentration of sugars build up, the concentration of acids decrease due, in part, to simple dilution but also to the consumption of acids in the process of plant respiration. The decrease in free acids, as well as the buildup of potassium, triggers a rise in the pH level of the grape juice.

In addition to the change in sugar, acids and pH levels other components of the grapes are building up during the ripening process. The mineral components of potassium, calcium, magnesium and sodium increase in concentration as they are disseminated among the skin of the grapes and its fleshy pulp. The color of the grape berries begin to change due to the building up of phenolic compounds such as anthocyanin in the skins. Flavonoids and volatile compounds known as "flavor precursors" which contribute to the eventual flavor and aroma of the wine also begin to build up in the skins and pulp. Additionally the concentration of tannins in the grape increases in several areas of the grape including the skin, seeds and stem. Early in the ripening process these tannins are very bitter and "green". Exposure to the warmth and sunlight during the ripening period ushers in chemical changes to the tannins that when processed into wine makes the tannins feel softer in the mouth.

Varying Ripeness Levels for Different Wines

Pinot noir grapes that are destined for sparkling wine will be considered ripe much earlier than Pinot noir destined for still red wine.

What constitutes "ripeness" will vary according to what style of wine is being produced as well as the particular views of winemakers and viticulturists on what optimal ripeness is. The style of wine is usually dictated by the balance between sugars and acids. What may be considered "ripe" for one winemaker could be considered under ripe to another winemaker or even overripe to yet a third winemaker. Climate and the particular grape variety will also play a role in determining ripeness and date of harvest. In very hot climates, such as certain areas in California and Australia, ripeness is usually achieved around 30 days after veraison starts while in much cooler climates, like the Loire Valley and parts of Germany, this may not occur until 70 days after veraison. The ripening periods for each individual grape variety will vary with grapes such as Cabernet Sauvignon taking much longer to ripen compared to early ripening varieties such as Chardonnay and Pinot noir.

Since over the course of ripening sugars in the grapes increase, the sweetness level as well as the potential alcohol level of the wine will play a considerable role in dictating when a grape is "ripe" enough. This is because sugars are converted by yeast into alcohol by the process of fermentation. The greater the concentration of sugars in the grape, the greater the potential alcohol level. However, most strains of winemaking yeast have difficulties surviving in an alcohol solution above 15% alcohol by volume (ABV) and cease fermentation before all the sugar is converted in alcohol. This leaves a certain amount of residual sugar which influences the sweetness level of the wine. Wines that are destined to be sweet, such as dessert wines, are often called late harvest wines because they are harvested at extreme points of ripeness much later than when regular table wine grapes have been harvested.

The presence of alcohol (particularly ethanol) in the wine contributes much more than just intoxication. It has an immense impact of the weight and mouthfeel of the wine as well as the balance of sweetness, tannins and acids. In wine tasting, the anaesthetic qualities of ethanol reduces the sensitivity of the palate to the harsh effects of acids and tannins, making the wine seem softer. It also plays a role during the ageing of wine in its complex interaction with esters and phenolic compounds that produce various aromas in wine that contribute to a wine's flavor profile. For this reason, some winemakers will value having a higher potential alcohol level and delay harvesting until the grapes have a sufficiently high concentration of sugars.

For other types of wines, such as sparkling wines like Champagne, maintaining a certain amount of acidity in the grapes is important to the winemaking process. As the concentration of acids in the grapes decrease the further along the ripening process you go, grapes destined for sparkling wines are often some of the earliest grapes to be harvested in a vintage. With their high acidity and low sugar levels, these grapes would be under ripe and would produce table wines that many wine drinkers would consider unpalatable yet the balance of sugars and acids is well suited for sparkling wine production.

Factors Influencing when Ripeness Occurs

One of the primary factors influencing the ripening process of grapevine is the climate and weather. Sunlight and temperature warmth are vital to the physiological functions of the grapevine (such as photosynthesis). An absence of either, such as long periods of extensive cloud cover, will cause many functions of the vine to slow or even completely halt as the vine enters a type of "survival mode". As the grapevine funnels more resources to preserve its own survival, less resources are directed towards the ripening and development of the grape clusters. Excessive heat can also cause a grapevine to react adversely. The occurrence of heat waves during the growing season, particularly as it nears harvest, can cause the sugars in grapes to jump as acids fall dramatically. Some winemakers may decide to harvest early in order to maintain acid levels even though other components (such as tannins and phenolic compounds) may not be at optimal ripening. For the winemakers that decide to "wait it out", a lack of acid can be partially rectify during the winemaking process with the addition of acids such as tartaric acid. It is much more difficult to remedy the effects of extensive rains during the ripening period. Steady rains before the harvest can cause the berries to swell with water which dilutes the flavors as well as causing cracking in the skin that creates openings for spoilage causing microorganism to propagate. Because of these risks, the threat of prolong rainfall during a vintage may cause an early harvest before the grapes have fully ripened. The most favorable vintages allow a slow, steady ripening without drastic jumps in heats or the threat of excessive rain fall.

Vineyard management techniques such as canopy management can influence the ripening process of grapes by balancing the amount of foliage needed for photosynthesis versus excessive foliage that shades the grapes and competes for the grapevine's resources.

The role that climate plays in influencing the ripening process cannot be overstated, but it is not the only factor. Vineyard management such as pruning and canopy management can also play a significant role as it not only influences the physiological processes of the grapevine but also how the vine responds in sharing its limited resources of energy and nutrients. The leaves of a grapevine produce energy via the process of photosynthesis. A certain amount of foliage is needed to ensure that the grapevine can produce enough energy to support all its physiological functions, but too much leaf cover will shade the grape clusters, limiting the direct exposure of sunlight and warmth needed for some chemical components of the grapes to develop. An excessive amount of foliage and shading may also promote the development of various vine diseases and ailments such as bunch rot and powdery mildew which can hamper the ripening process. A very vigorous vine with many clusters and vine shoots will have several parties competing for the same resources, with the overall development of an individual clusters thus slowed. Through the process of canopy management, viticulturists try to balance not only the amount of clusters and vine shoots on the vine but also try to achieve an optimal balance of needed foliage for photosynthesis without excessive shading that could hamper the ripening process.

Even if climate and vineyard management has been ideal, other factors may prevent full and even ripeness. Among the clusters of a grapevine, individual berries may not all ripen at the same pace. This problem, commonly known as millerandage, could occur because of poor weather during the flowering period of the grape but can also be caused by soil deficient in various nutrients such as boron, an attack of various grapevine ailments such as the grapevine fanleaf virus or a number of other factors that may contribute to incomplete plant fertilization.

Evaluating Ripeness

As "ripeness" constitutes a variety of factors, there are many methods that viticulturist and winemakers may use in order to determine when the grapes are sufficiently ripe to harvest. The most common method of determining ripeness involves measuring the sugar, acid and pH levels of the grapes with the purpose of harvesting at point when each number reaches its most ideal range for the type of wine being produced. In recent years, viticulturists and winemakers have shifted away from focusing purely on those numbers towards considering other factors including the ripeness of tannins, the development of flavor precursors and the potential for glycosides to development. A combination of these factors apart from sugar, acid and pH are considered "physiological" ripeness of the grape.

Grapes that have been left on the vine too long may become over ripe and dehydrated.

Must Weight

Since more than 90% of all the dissolved solids in grape juice are sugars, measuring the must weight is a good indicator of the amount of sugars in the wine. Rather than measure the actual "weight" of the must, the density or specific gravity of the juice is measured in relation to the specific gravity of distilled water. Viticulturists and winemakers can use a refractometer which uses a refractive index to indirectly measure the must weight from the juice of a single grape or they can use a hydrometer in the winery with the juice from several dozen or hundreds of grape berries. Different countries around the world use various scales to measure the must weight of grape juice. In the United States, New Zealand, and parts of Australia it is measured in degrees brix (symbol °Bx); in Germany (wine) it is degrees Oechsle (°Oe); in France and most of Europe the Baumé scale was used until 1961 and in Austria the Klosterneuburger Mostwaage (°KMW) scale is used.

After veraison has begun, viticulturists will test several hundred individual berries picked from clusters throughout the vineyard in increasing intervals as the harvest draws closers. The berries will usually be taken from the middle of the cluster bunch, avoiding vines on the end of rows that tend to be exposed to the most unusual elements. The must weight is then plotted on a chart to see the increasing ripeness and sugar levels of the grape. What must weight reading is most desirable will depend on the winemaker's personal goal for ripeness. A wine with the intended potential alcohol level of 12% will need to be harvested at around 21.7°Bx/12 degree Baumé/93°Oe. A wine with the intended potential alcohol level of 15% will need to be harvested at around 27.1°Bx/15 degree Baumé/119°Oe. The desired ripeness for most table wines tend to fall somewhere between those two must weight measurements.

Acid Level

As sugar levels in the grape rise, acid levels fall. All wines need some degree of acidity in order to be balanced and avoid tasting flabby or dull. Acidity is also a key component in food and wine pairing so its presence in wine is important with winemakers trying to harvest grapes before acid levels fall too low. The stress to maintain acid levels is not as bearing due to the fact that winemakers can rectify the situation somewhat by later adding acids during the winemaking process (winemakers can also rectify deficiencies in sugar levels by chaptalization). However natural acids in the grape play other roles in the development of flavor and aroma compounds as well as fighting against the effects of spoilage organisms so the most ideal situation for winemakers is to try and harvest while acid levels are acceptable.

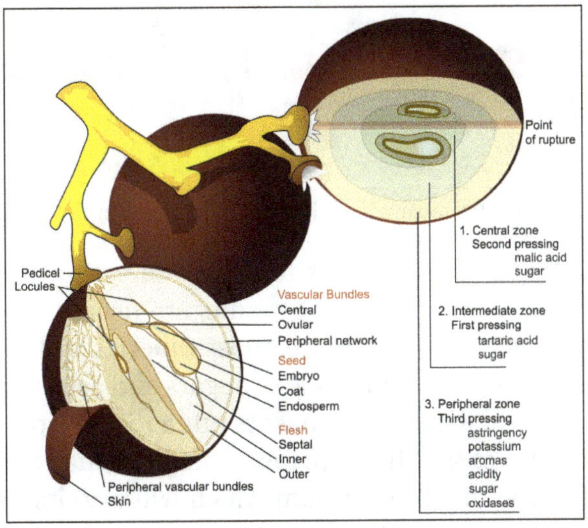

The principle acids found in wine grapes are tartaric and malic acids.

The major acids in wine are tartaric and malic acid with citric and succinic acids playing a small role. The titratable acidity or "TA" (also referred to as "total acidity") is the measure of the tartaric acid in the grapes. This is the most abundant acid and also the one acid that has the most pronounced and long lasting impact on the taste of the wine. The TA is often measured by neutralizing some grape juice with a standard alkaline solution (such as sodium hydroxide) and then using an indicator (such as phenolphthalein) which changes color depending on the acid levels of the solution. The indicator is added to the grape juice followed by incremental amounts of the alkaline solution as the wine changes color until adding more of the solution ceases to promote a color change. At this point the wine has been neutralized with the amount of the alkaline solution needed to neutralize calculated in a formula to give an indication of how much tartaric acid was in the wine. The TA level is then expressed in a percentage of grams per 100 milliliter. As with must weight, the ideal levels for ripeness will vary according to wine style and winemaking preference. For still table wines, TA levels often fall between 0.60-0.80% for red wine grapes and 0.65-0.85 for whites.

pH Level

The pH level of a wine is the measurement of the amount of free (H+) hydrogen ions. It is related to the titratable acidity level of a wine but differs in significant ways. Low pH numbers indicate a high concentration of acids in a solution. While pure water is neutral with a pH of 7, wine tends to be more acidic with a pH between 3 and 4. As the acid levels in ripening grapes fall, the concentration of acids are lessening which means the pH level is rising. Yeasts, bacteria, phenolic compounds such as anthocyanins which influence color all have varying tolerance to wines with high pH levels.

In general, wines with high pH value tend to have duller colors and less developed flavors and be more prone to wine faults caused by spoilage organisms which makes monitoring the pH levels of grapes during ripening a priority for viticulturists and winemakers.

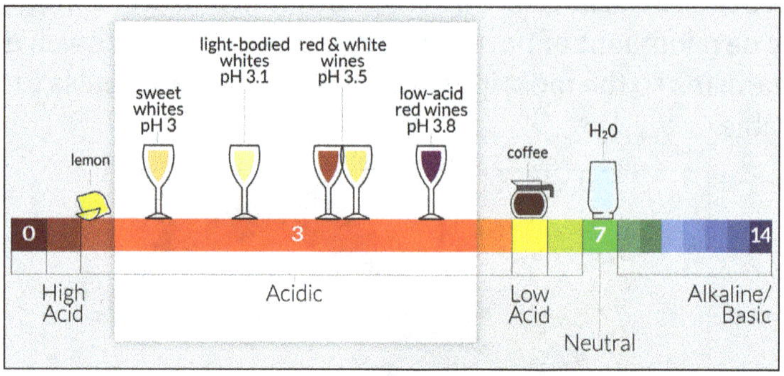

The pH levels for most wine fall between 3 and 4 on the pH scale.

While the rudimentary method of testing pH is to expose the grape juice to a pH indicator such as the strips used for a standard litmus test, the results are usually not as detailed and accurate as what is needed to evaluate ripeness. Therefore, most wineries will us a pH meter that can give readings to an accuracy of plus or minus 0.1. As with sugars and acids, the ideal pH levels to determine ripeness will vary. For white wines, winemakers often look for pH readings between 3.1 and 3.2, while would be a maximum of 3.4. If the pH is too high, it may be a sign that the grapes are overripe (or that the soil has too much potassium which will also influence pH readings). While there are risks to letting the pH go too high, winemakers can counter high pH by adding more tartaric or malic acid during the winemaking. However many viticulturists and winemakers uses pH readings as a strong boundary line for when to start the harvest.

Balancing Sugar, Acidity and pH

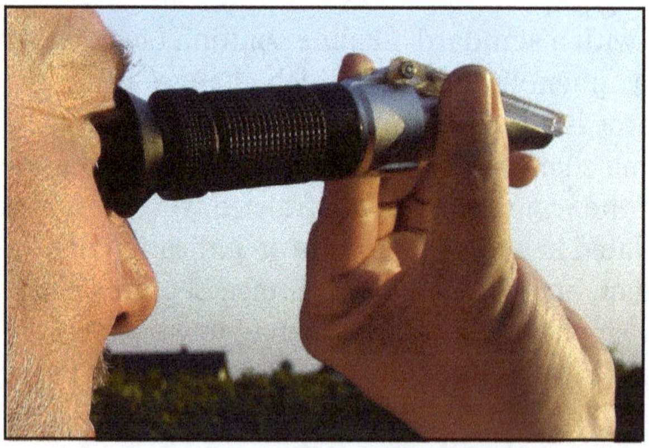

Winemakers use a refractometer on samples of grapes picked in the vineyard to measure sugar levels while determining ripeness.

The most ideal situation for a viticulturist or winemaker is to have the sugar, acidity and pH levels to be perfectly balanced at the time of harvesting. One hypothetical ideal for still red table wine is to have grape measurements reading 22 Brix, 0.75 TA and 3.4 pH. As author and winemaker Jeff Cox notes, these numbers are the "royal flush" poker hand of winemaking that is rarely dealt to winemakers. With all the variables of climate, vineyard soils, grape varieties, vineyard management and the general characteristics of the vintage, winemakers learn to find a compromise between all these component readings and select the point of ripeness that is most align with their vision for the end product wine.

There are several formulas that viticulturist and winemakers can use that utilize the various measurements of sugar, acid and pH level. One method developed by researchers at the University of California-Davis is the Brix:TA ratio which uses the ratio of brix degrees to the TA measurements. For example, a wine with 22°Bx and 0.75 TA will have almost a 30:1 Brix:TA ratio. According to the Davis researchers, the most balanced table wines tend to have a Brix to TA ratio between 30:1 - 35:1. Another method is to multiply the pH reading by itself and then multiply that number by the Brix reading. Using this method, when white wine grapes gets close to 200 and red wine grapes close to 260, it can be a good rule of thumb of when to harvest. For example, white wine grapes have a pH of 3.3 and Brix of 20, after going through that formula they will have a finally number of 217.80 which is well within an acceptable harvest range for some winemakers.

Physiological Ripeness

The idea of physiological ripeness (or physiological maturity) of grapes is a relatively recent addition to the discussion of ripeness in viticulture and winemaking. It is a broad category of factors in the development of ripening grapes that affect a wine's quality beyond the standard measurements of sugars, acids, and pH. These factors generally include evaluating the ripeness of tannins as well as the development of other phenolic compounds that contribute to the color, flavor, and aroma of wine. In many ways, the concept of physiological ripeness is similar to the French notion of engustment, the stage of ripening when aroma and flavor become apparent. Research has shown that most aroma compounds develop in the berry in glycosylated form as secondary metabolites which occur late in ripening as the buildup of sugars have leveled. This stage is distinct from the sugar/acid interactions of ripening because it is possible for a grape to be "ripe" in the context of sugar and acid levels but still be very immature when it comes to the development of tannins, aromas and flavor that characteristic a complex or quality wine.

In determining physiological ripeness, winemakers will observe the lignification of the grape stems as they turn from being flexible and green to hard, woody and brown.

For the most part, many of these qualities are difficult to objectively measure so evaluation of the physiological ripeness of grapes is centered around observing and physically sampling the grapes. With experience winemakers and viticulturists learn to associate certain taste and characteristics with different stages of development. They evaluate the skin and pulp texture of the berry as well as the color of skins, seeds and stems. If the seeds are still green, the tannins inside the grape are more likely to be harsh and bitter. As the tannins continue to develop, the seeds start darkening in color. They will observe the lignification of the stems as they turn from being flexible and green to hard, woody and brown (for many varieties but not all) indicating that vine has completed its work in developing its "offspring" grape clusters and has started to store carbohydrates and resources for its next growing season. During the ripening period winemakers and viticulturists will continually sample grapes throughout the vineyard in the weeks and days leading up to harvest.

Flavor Precursors and Glycosides

While it is difficult to objectively measure the qualities of physiological ripeness, researchers in the wine industry have been continuing pursuing methods that give some indication of the grapes development in these areas. For instance, some wineries have started using near infrared (NIR) spectroscopy to determine the concentration of color producing anthocyanins in the skins of grapes. A sizable amount of research has gone into studying methods to determine the presence of flavor precursors and glycosides in the ripening grapes.

Researchers in the wine industry are developing new ways to objectively measure ripeness.

Recently, similar methods to determine chlorophyll content in leaves non-destructively have been applied to measuring anthocyanin content. There are now a couple of optical absorbance instruments available commercially which are designed to measure and compute an index value that correlates highly with the actual amount of anthocyanin content in a sample. To use with grapes, the skin is removed and placed across the sensor of the meter. Measurements take only a second or two. These Anthocyanin Content Meters use an additional Near Infra-Red (NIR) signal, which takes into account the thickness of the sample, along with the absorbance wavelength to calculate a very accurate index value which is repeatable and consistent enough for comparative testing. A new method just being explored is to dip a piece of filter paper into a solution/sample to be measured and put that across the sensor head as the test sample.

Flavor precursors are flavorless compounds that occur naturally in grapes as the result of normal metabolic activity of the grape vine. They are more abundant in grapes than the phenolic compounds known as flavonoids, and include compounds such monoterpenes, which contributes to the floral aroma of Riesling and Muscat, and methoxypyrazine, which contributes to the "green-bell pepper" aroma associated with Cabernet Sauvignon and Sauvignon blanc. When these components are "free" they are known as "flavor compounds" but when they combine with sugars in the grapes, they become glycosides or "flavor precursors". These compounds are found in trace amounts, and measured in parts per trillion. Through the action of acids and enzymes, glucosides derived from the sugar in the grapes go through hydrolysis, creating glycosides. These compounds are released during the late stages of winemaking and aging, when they augment or enhance flavor compounds. Theoretically, grapes with more flavor precursors have the potential to produce higher quality wine.

Scientist have discovered it is possible to determine, to some extent, the presence of these compounds in the grape before harvest. One way is to measured with gas chromatograph-mass spectrometers. Another method is through analysis of the glycosyl-glucose assay. Through this method glycosides from the grape juice are isolated and then hydrolysized to yield glucose. The amount of glucose produced is then quantified and tabulated in results that are expressed as amount of glycosides in micromoles per liter or per grape berry. The relationship between the presence of glycosides in wine grapes and the potential for quality in the resulting wine is not exact science but this remains an area of continuing research and development.

Urban Horticulture

Urban horticulture specifically is the study of the relationship between plants and the urban environment. It focuses on the functional use of horticulture so as to maintain and improve the surrounding urban area. With the expansion of cities and rapid urbanization, this field of study is large and complex and its study has only recently gained momentum. It has an undeniable relationship to production horticulture in that fruits, vegetables and other plants are grown for harvest, aesthetic, architectural, recreational and psychological purposes, but it extends far beyond these benefits. The value of plants in the urban environment has yet to be thoroughly researched or quantified.

Horticulture and the integration of nature into our civilization has been a major part in the establishment of our cities. When nomadic civilizations began settling down, their major trading centers were the market gardens and farms. Urban horticulture rapidly progressed with the birth of cities and the increase in experimentation and exchange of ideas. These insights led to the field being dispersed to farmers in the hinterlands. For centuries, the built environment such as homes, public buildings, etc. were integrated with cultivation in the form of gardens, farms, and grazing lands, Kitchen gardens, farms, common grazing land, etc. Therefore, horticulture was a regular part of everyday life in the city. With the Industrial Revolution and the related increasing populations rapidly changed the landscape and replaced green spaces with brick and asphalt. After the nineteenth century, Horticulture was then selectively restored in some urban spaces as a response to the unhealthy conditions of factory neighborhoods and cities began seeing the development of parks.

Post World War II Trends

Early urban horticulture movements majorly served the purposes of short term welfare during recession periods, philanthropic charity to uplift "the masses" or patriotic relief. The tradition of urban horticulture mostly declined after World War II as suburbs became the focus of residential and commercial growth. Most of the economically stable population moved out of the cities into the suburbs, leaving only slums and ghettos at the city centers. However, there were a few exceptions of garden projects initiated by public housing authorities in the 1950s and 1960s for the purpose of beautification and tenant pride. But for the most part as businesses also left the metropolitan areas, it generated wastelands and areas of segregated poverty.

Inevitably the disinvestment of major city centers, specifically in America resulted in the drastic increase of vacant lots. Existing buildings became uninhabitable, houses were abandoned and even productive industrial land became vacant. Modern community gardening, urban agriculture, and food security movements were a form of response to battle the above problems at a local level. In fact other movements at that time such as the peace, environmental, women's, civil rights, and "back-to-the-city" movements of the 1960s and 1970s and the environmental justice movement of the 1980s and 1990s saw opportunity in these vacant lands as a way of reviving communities through school and community gardens, farmers' markets, and urban agriculture.

Modern Community Garden Movement

Things have taken a turn in the twenty-first century as people are recognizing the need for local community gardens and green spaces. It is not the concept but the purposes that are new. The main goals of this movement include cleaning up neighborhoods, pushing out drug dealing that occurs at empty lots, growing and preserving food for consumption, restoring nature to industrial areas, and bringing the farming traditions to urban cities. Essentially community gardening is seen as way of creating a relationship between people and a place through social and physical engagement. Most urban gardens are created on vacant land that vary in size and are generally gardened as individual plots by community members. Such areas can support social, cultural, and artistic events and contribute to the rebuilding of local community spirit. The modern community garden movement is initiated by neighborhoods along with the support of the governments and non-profit organizations. Some gardens are linked to public housing projects, schools through garden-based learning programs, churches and social agencies and some even employ those who are incarcerated. Community gardens which are now a large part of the urban horticulture movement are different from the earlier periods of grand park development in that the latter only served to free the people from the industrialism. In addition a community garden is more beneficial and engaging than a mere lawn or park and serves as a valuable access to nature where wilderness is unavailable. This movement helped create and sustain relationships between city dwellers and the soil and contributed to a different kind of urban environmentalism that did not have any characteristics of reform charity.

Despite that it has been 30 years since the first community gardens in the US, there is no concrete analysis of current urban gardens and their organizations. The American Community Gardening Association (ACGA) has estimations that show that municipal governments and non-profit organizations operate gardening programs in about 250 cities and towns, although the staff of this organization admits that this number could in reality be twice as large. In 1994 survey, the National Gardening Association found that 6.7 million households, that weren't involved in gardening would be interested in doing so if there was a plot nearby. A more recent survey showed that more gardens are being created in cities as opposed to being lost to economic development.

Today urban horticulture has several components that include more than just community gardens, such as market gardens, small farms and farmers' markets and is an important aspect of community development. Another result of urban horticulture is the food security movement where locally grown food is given precedence through several projects and programs, thus providing low-cost and nutritious food. Urban community gardens and the food security movement was a response to the problems of industrial agriculture and to solve its related problems of price inflation, lack of supermarkets, food scarcity, etc.

Benefits

Horticulture by itself is a practical and applied science, which means it can have a significance in our everyday lives. As community gardens cannot actually compete with market-based land uses, it is essential to find other ways to understand their various benefits such as their contribution to social, human, and financial well-being. Frederick Law Olmsted, the designer of New York City's Central Park observed that the trees, meadows, ponds and wildlife tranquilize the stresses of city life. According to various studies over the years, nature has a very positive impact over human health and even more so in an emotional and psychological sense. Trees, grass, and flower gardens, due to their presence as well as visibility, increase people's life satisfaction by reducing fatigue and irritation and restoring a sense of calm. In fact Honeyman tested the restorative value of nature scenes in urban settings and discovered that vegetation in an urban setting produced more mental restoration as opposed to areas without vegetation. In addition, areas with only nature did not have as much of a positive psychological impact as did the combination of urban areas and nature.

One of the obvious health benefits of gardening is the increased intake of fruits and vegetables. But the act of gardening itself, is also a major health benefit. Gardening is a low-impact exercise, which when added into daily activities, can help reduce weight, lower stress, and improve overall health. A recent study showed a reduced body mass index and lower weight in community gardeners compared with their non-gardening counterparts The study showed men who gardened had a body mass index 2.36 lower and were 62% less likely to be overweight than their neighbors, while women were 46% less likely to be overweight with a body mass index 1.88 lower than their neighbors. Access to urban gardens can improve health through nutritious, edible plantings, as well by getting people outside and promoting more activity in their environments.

Gardening programs in inner-city schools have become increasingly popular as a way to teach children not only about healthy eating habits, but also to encourage students to become active learners. Besides getting students outside and moving, and encouraging an active lifestyle, children also learn leadership, teamwork, communication and collaboration skills, in addition to critical and creative thinking skills. Gardening in schools will enable children to share with their families the health and nutrition benefits of eating fresh fruits and vegetables. Because weather and soil conditions are in a state of constant change, students learn to adapt their thinking and creatively problem solve, depending on the situations that arise. Students also learn to interact and communicate with a diverse population of people, from other students to adult volunteers. These programs benefit students' health and enable them to be active contributors in the world around them.

Gardens and other green spaces also increase social activity and help in creating a sense of place, apart from their various other purposes such as enhancing the community by mediating environmental factors. There is also a huge disparity in the availability of sources that provide nutritious and affordable foods especially around urban centers which have problems of poverty, lack of public transport and abandonment by supermarkets. Therefore, inner city community gardens can be a valuable sources of nutrition at an affordable cost in the most easily accessible way.

In order to understand and thereby maximize the benefits of urban horticulture, it is essential to document the effects of horticulture activities and quantify the benefits so that governments and private industries can make the appropriate changes. Horticulturists have always been involved in the botanical and physical aspects of horticulture but an involvement in its social and emotional factors would be highly beneficial to communities, cities and to the field of horticulture and its profession. Based on this, in the 1970s, the International Society for Horticultural Science recognized this need for research on the functional use of plants in an urban setting along with the need of improved communication between scientists in this field of research and people who utilize plants. The Commission for Urban Horticulture was established in 1982 which deals with plants grown in urban areas, management techniques, the functional use of these plants as well the shortcomings of the current lack of knowledge regarding this field. The establishment of such a commission is an important indicator that this topic has reached a level of international recognition.

Economic Benefits

There are many different economic benefits from gardening from saving money purchasing food and even on the utility bills. Developing countries can spend up to 60–80 percent of income on buying food alone. Having green roofs can reduce the cost of heating in the winter and help stay cool in the summer. Green roofs also can lower the cost of roof replacement. While green roofs are an addition to urban horticulture people are eating healthy while also improving the value of their property. Other benefits include increased employment from non-commercial jobs where producers include reductions on the cost of food.

Production Practices

Tomato plants growing in a pot farming alongside a small house in New Jersey in fifteen garbage cans filled with soil, grew over 700 tomatoes during the summer.

Crops are grown in flowerpots, growbags, small gardens or larger fields, using traditional or high-tech and innovative practices. Some new techniques that have been adapted to the urban situation and tackle the main city restrictions are also documented. These include horticultural production on built-up land using various types of substrates (e.g. roof top, organic production and hydroponic/aeroponic production). Because of this, it is also known as roof-top vegetable gardening/horticulture and container vegetable gardening/horticulture.

Arboriculture

Arboriculture is the cultivation of trees, shrubs, and woody plants for shading and decorating. Arboriculture includes propagating, transplanting, pruning, applying fertilizer, spraying to control insects and diseases, cabling and bracing, treating cavities, identifying plants, diagnosing and treating tree damage and ailments, arranging plantings for their ornamental values, and removing trees. The well-being of individual plants is the major concern of arboriculture, in contrast to such related fields as silviculture and agriculture, in which the major concern is the welfare of a large group of plants as a whole.

The basic principles and objectives of arboriculture are of ancient origin. Early Egyptians transplanted trees with a ball of earth and originated the practice of shaping the soil around a newly planted tree to form a saucer to retain water, both still practiced. About 300 BC the Greek philosopher Theophrastus wrote Peri phytōn historia ("Inquiry into Plants"), in which he discussed transplanting of trees and the treatment of tree wounds. Virgil's Georgics portrays Roman knowledge of tree culture. The English horticulturist John Evelyn, in his Sylva, or a Discourse of Forest-trees, and the Propagation of Timber, offered advice on pruning, insect control, wound treatment, and transplanting.

Trees or plants may be propagated by seeding, grafting, layering, or cutting. In seeding, seeds are usually planted in either a commercial or home nursery in which intensive care can be given for several years until the plants are of a size suitable for transplanting on the desired site. In soil layering, the shoots, or lower branches of the parent plant, are bent to the ground and covered with moist soil of good quality. When roots have developed, which may require a year or more, the branch is severed from the parent and transplanted. In an alternative technique, air layering, the branch is deeply slit and the wound covered by a ball of earth, moss, or similar material. The ball, enclosed in a divided pot supported from underneath, or in a sturdy paper cone, is kept moist. As in soil layering, the branch is severed and transplanted after roots have developed. Root cuttings can be used for propagating trees that do not normally produce roots from stems. Tree species such as willow and poplar that sucker, or send up shoots readily, are usually propagated from stem cuttings. Cuttings are made from deciduous plants during dormancy, preferably from the terminal growing shoots of the current season. Pieces 6 to 10 inches (15 to 25 centimetres) long with two or more buds are tied in bundles and stored in damp sand or moss for callus formation before planting in prepared beds. Root formation may be stimulated by application of growth-promoting chemicals or growth hormones.

In treating tree-trunk wounds in which large areas of bark are torn away, the bark around the wound is trimmed back to sound tissue and, at the top and bottom of the injury, trimmed to form a pointed ellipse of the wound area. The exposed wood is covered with wound dressing material, protecting it from wood decay fungi.

Flexible cables (guys) or rigid braces are used to support recently transplanted trees until the roots become established, or to lessen the danger that a tree with a weakened root system will be blown over by the wind; bracing is also used to support unduly long or heavy branches, to prevent splits developing at branch forks, or to permit healing of splits already developed.

Cavities in trunks, caused by decay-inducing fungi, may be treated with antiseptic dressing and left open, with drains installed at the bottom, or filled with concrete or other material after removal of the decayed wood.

Tree Shaping

Tree shaping (also known by several other alternative names) uses living trees and other woody plants as the medium to create structures and art. There are a few different methods used by the various artists to shape their trees, which share a common heritage with other artistic horticultural and agricultural practices, such as pleaching, bonsai, espalier, and topiary, and employing some similar techniques. Most artists use grafting to deliberately induce the inosculation of living trunks, branches, and roots, into artistic designs or functional structures.

Tree shaping has been practiced for at least several hundred years, as demonstrated by the living root bridges built and maintained by the Khasi people of India. Early 20th century practitioners and artisans included banker John Krubsack, Axel Erlandson with his famous circus trees, and landscape engineer Arthur Wiechula. Contemporary designers include "Pooktre" artists Peter Cook and Becky Northey, "arborsculpture" artist Richard Reames, and furniture designer Dr Chris Cattle, who grows "grownup furniture".

Methods

There are various methods of shaping a tree. Some of these processes are still experimental, whereas others are still in the research stage. These methods use a variety of horticultural and arboricultural techniques to achieve an intended design. Chairs, tables, living spaces and art may be shaped from growing trees. Some techniques used are unique to a particular practice, whereas other techniques are common to all, though the implementation may be for different reasons. These methods usually start with an idea of the intended outcome. Some practitioners start with detailed drawings or designs. Other artists start with what the tree already has. Each process has its own time frame and a different level of involvement from the tree shaper. The trees might then either remain growing, as with the living Pooktre garden chair, or perhaps be harvested as a finished work, like John Krubsack's chair.

Aeroponic Culture

The oldest known living examples of woody plant shaping are the aeroponically cultured living root bridges built by the ancient War-Khasi people of the Cherrapunjee region in India. These are being maintained and further developed today by the people of that region. Aeroponic growing was first formally studied by W. Carter in 1942. Carter researched air culture growing and described "a method of growing plants in water vapor to facilitate examination of roots". Later researchers, including L. J Klotz and G. G. Trowel, expanded on his work. In 1957, F. W. Went described "the process of growing plants with air-suspended roots and applying a nutrient mist to the root section", and in it he coined the word 'aeroponics' to describe that process. In 2008, root researcher and craftsman Ezekiel Golan described and secured a patent for a process which allows the roots of some aeroponically grown woody plants to lengthen and thicken while still remaining flexible. At lengths of perhaps 6 metres (20 ft) or more, the soft roots can be formed into pre-determined shapes which will continue thickening after the shapes are formed and as they continue to grow. Newer techniques and applications, such as eco-architecture, may allow architects to design, grow, and form large permanent structures, such as homes, by shaping aeroponically grown plants and their roots.

Instant Tree Shaping

Instant tree shaping starts with mature trees, perhaps 6–12 ft. (2–3.6 m) long and 3-4in (7.6–10 cm) in trunk diameter, which are bent and woven into the desired design and held until cast. Understanding a tree's fluid dynamics is important to achieving the desired result.

Bending is sometimes used to achieve a design. If a plant's tissue is bent at too sharp an angle it may break, which can be mostly avoided by un-localizing the bend. This is achieved by making small bends along the curve of the tree. Bends are then held in place for several years until their form is permanently cast. The tree's rate of growth determines the time necessary to overcome its resistance to the initial bending. The work of bending and securing in this way might be accomplished in an hour or perhaps in an afternoon depending on the design.

Ring barking is sometimes employed to help balance a design by slowing the growth of too-vigorous branches or stopping the growth of inopportunely placed branches, using different degrees of ring barking, from simple scoring to complete removal of a 3/8"-wide (1 cm) band of bark.

Creasing is folding trees such as willow and poplar over upon themselves, creating a right angle. This method is more radical than bending.

With this method it is possible to perform initial bending and grafting on a project in an hour, as with Peace in Cherry by Richard Reames, removing supports in as little as a year and following up with minimal pruning thereafter.

Gradual Tree Shaping

Gradual tree shaping starts with designing and framing. These are fundamental to the success of the piece. Once these are set up, young seedlings or saplings 3–12 in. (7.6–30.5 cm) long are planted.

The training starts with young seedlings, saplings or the stems of trees when they are very young, which are gradually shaped while the tree is growing to form the desired shape. There is a small area just behind the growing tip that forms the final shape. The shaping zone, it is the shaping of this area requires day to day or weekly guiding of the new growth. The growth is guided along predetermined design pathways; this may be a wooden jig or complex wire design.

With this method the time frame is longer than the other methods. A chair design might take 8 to 10 years to reach maturity Some of Axel Erlandson trees's took as long as 40 years to assume their finished shapes.

Common Techniques

Framing

Framing may be used for various purposes and might consist of any one or a combination of several materials, such as timber, steel, hollowed out trees, complex wire designs, wooden jigs, or the tree itself, living or dead. It can be used in many project designs to support grafted joints until the grafts are well-established. Some processes might employ framing to hold a shape created by bending or fletching mature trees until the tissues have overcome their resistance to the initial bending and grown enough annual rings to cast the design permanently. Others might use framing to support and shape the growth of young saplings until they are strong enough to maintain an intended shape without support. Still other approaches might employ frames to guide the roots of aeroponically grown trees into desired shapes.

Grafting

Grafting is a commonly employed technique that exploits the natural biological process of inosculation. A branch or plant is cut and a piece of another plant is added and held in place. Various types of grafting all share the goal of encouraging the tissues of one plant to fuse with those of another.

Grafting is applied to create permanent connections and joints. In some cases, trees are grafted while they are growing, while in other cases, mature trees may be intertwined and the stems of two or more trees are then grafted together to create chairs, ladders, and other fanciful sculptures.

Pruning

Pruning can be used to balance a design by controlling and directing growth into a desired shape. Pruning above a leaf node can steer plant growth in the direction of the natural placement of that leaf bud. Pruning may also be used to keep a design free of unwanted branches and to reduce canopy size. Pruning is sometimes the only technique used to craft a project. Deciduous trees are mainly pruned in winter, while they are dormant above-ground, although sometimes it is necessary to prune them during the growing season. Trees repeatedly subjected to hard pruning may experience stunted growth, and some trees may not survive this treatment.

Structure

Living grown structures have a number of structural mechanical advantages over those constructed of lumber and are more resistant to decay. While there are some decay organisms that can rot live wood from the outside, and though living trees can carry decayed and decaying heartwood inside them; in general, living trees decay from the inside out and dead wood decays from the outside in. Living wood tissue, particularly sapwood, wields a very potent defense against decay from either direction, known as compartmentalization. This protection applies to living trees only and varies among species.

Growing structures is not as easy as it would seem. Quick growing willows have been used to grow building structures, they provide support or protection. A young group of German architects are in the process of such a structure and they are continually monitored and checked. Once the trees are of age to be able to take on load-bearing weight they are tested for stability and strength by a structural engineer. Once this is approved the supporting framework is removed. Projects are limited to the trees' weight loading ability and growth. This is being studied and the load capacity will be proved by testing on prototypes.

Design options

Designs may include abstract, symbolic, or functional elements. Some shapes crafted and grown are purely artistic; perhaps cubes, circles, or letters of an alphabet, while other designs might yield any of a wide variety of useful shapes, such as clothes hangers, laundry and wastepaper bins, ladders, furniture, tools, and tool handles. Eye-catching structures such as living fences and jungle gyms can also be grown, and even large architectural designs such as live archways, domes, gazebos, tunnels, and theoretically entire homes are possible with careful planning, planting, and culturing over time. The Human Ecology Design team (H.E.D.) at the Massachusetts Institute of Technology is designing homes that can be grown from native trees in a variety of climates.

Becky's Mirror by Pooktre.

Suitable trees are installed according to design specifications and then cultured over time into intended structures. Some designs may use only living, growing wood to form the structures, while others might also incorporate inclusions such as glass, mirror, steel and stone, any of which might be used either as either structural or aesthetic elements. Inclusions can be positioned in a project as it is grown and, depending on the design, may either be removed when no longer needed for support or left in place to become fixed inclusions in the growing tissue.

Related Practices

Other artistic horticultural practices such as bonsai, espalier, and topiary share some elements and a common heritage, though a number of distinctions may be identified.

Bonsai

Bonsai is the art of growing trees in small containers. Bonsai uses techniques such as pruning, root reduction, and shaping branches and roots to produce small trees that mimic full-sized mature trees. Bonsai is not intended for production of food, but instead mainly for contemplation by viewers, like most fine art.

Espalier

Espalier is the art and horticultural practice of training tree branches onto ornamental shapes along a frame for ascetic and fruit production by grafting, shaping and pruning the branches so that they grow flat, frequently in formal patterns, against a structure such as a wall, fence, or trellis. The practice is commonly used to accelerate and increase production in fruit-bearing trees and also to decorate flat exterior walls while conserving space.

Pleaching

Pleaching is a technique of weaving the branches of trees into a hedge commonly, deciduous trees are planted in lines, then pleached to form a flat plane on clear stems above the ground level. Branches are woven together and lightly tied. Branches in close contact may grow together, due to a natural phenomenon called inosculation, a natural graft. Pleach also means weaving of thin, whippy stems of trees to form a basketry affect.

Topiary

Topiary is the horticultural practice of shaping live trees, by clipping the foliage and twigs of trees and shrubs to develop and maintain clearly defined shapes, often geometric or fanciful. The hedge is a simple form of topiary used to create boundaries, walls or screens. Topiary always involves regular shearing and shaping of foliage to maintain the shape.

Plantings for the Future

Fab Tree Hab

Fab Tree Hab 3D render.

Three MIT designers – Mitchell Joachim, Lara Greden and Javier Arbona – created a concept of a living tree house which nourishes its inhabitants and merges with its environment. The project of Fab Tree Hab is expected to take a minimum of five years to grow the home. The plans are for the interior to be lined with clay and plastered to keep the weather outside and to look normal. The exterior is to be all natural.

Patient Gardener

A Swedish architectural firm VisionDivision took part in a week-long workshop at the Italian university Politecnico di Milano with the students. The result was an 80-year plan of a living cherry tree dome in an hourglass shape and grown furniture. Framing for the dome, table and a lawn chair were built. Ten Japanese cherry trees were planted in a diameter of eight-meter circle. Four of these trees are to be living staircases to a future top level. The stair trees will have their branches grafted into each other to form the rungs. Vision Division's architects helped the students and instructors to create an easy maintenance plan for future gardeners of the university.

Baubotanik Tower

Ferdinand Ludwig designed this tower as part of his doctoral thesis with the help of Prof. Dr. Speck. "Speck become the botanical co-supervisor" said Ferdinand. Growing at the University of Stuttgart is a three-storey tower of living white willows (Salix alba). This nine-meter-tall construction is almost fully grown, with a base area of around eight square meters.

The framing is made up of mainly steel scaffolding which is supporting the growing trees, while keeping them to the correct form. They started with 400 white willow (Salix alba) grown in baskets on multiple levels with one row of willows planted into the ground. Once the trees were two meters tall, they were planted at the different levels of the tower. These plants are then trained to the design.

The root system of the bottom level of willows needs to develop large enough to support the willows on the above levels, so that the scaffold becomes obsolete and then it and the watering and fertilising baskets can be removed altogether.

The trees are grafted together with the objective of all the different plants eventually becoming a single organism. The overall aim is to have a living structure with the strength to support itself and to carry a working load. Ferdinand predicts the tower will be stable enough to support itself in five to ten years.

References

- Horticulture, science: britannica.com, Retrieved 6 January, 2019
- Floriculture, farm-based-enterprises, agriculture: vikaspedia.in, Retrieved 13 May, 2019
- Proffitt, t., r. Bramley, d. Lamb, and e. Winter. 2006. Precision viticulture: a new era in vineyard management and wine production. Winetitles, adelaide. Isbn 978-0-9756850-4-4
- What-is-floriculture-and-its-scope: blogspot.com, Retrieved 8 January, 2019
- J. Robinson (ed) "the oxford companion to wine" third edition pg 255-274, 317-324, 397, 523-524, 582-581 oxford university press 2006 isbn 0-19-860990-6
- Fruit-farming, topic: britannica.com, Retrieved 21 May, 2019
- Goode, j. 2005. The science of wine: from vine to glass. University of california press, berkeley. Isbn 0-520-24800-7, isbn 978-0-520-24800-7
- What-is-viticulture: worldatlas.com, Retrieved 26 July, 2019
- Relf, diane. "human issues in horticulture". Virginia tech. Department of horticulture. Retrieved 6 october 2014

Diverse Aspects of Horticulture

2

- Ornamental Plants
- Plug
- Bedding
- Bokashi
- Bolting

A few of the fundamental concepts of horticulture are plug, bedding, bokashi and bolting. It also includes plant development, plant growth and plant embryogenesis. This chapter closely examines these fundamental concepts of horticulture to provide an extensive understanding of the subject.

Ornamental Plants

Ornamental plants are grown for use by the green industry and public for purposes such as landscaping for sport, and conservation. The green industries include commercial plant nurseries, flower growers, parks, and roadside and landscape plant installation and maintenance.

The primary use for these plants is not for food, fuel, fiber, or medicine. However, ornamental plants contribute significantly to the quality of life by acting as barriers to wind, providing cooling shade, reducing or eliminating erosion, cleaning the air and water of pollutants including dust and chemicals, reducing noise pollution, and providing food and habitat for wildlife while making both suburban and urban areas more beautiful. Their economic and emotional impact is significant.

Ornamental plants include perennial deciduous and evergreen shade trees, conifers, and shrubs grown in horticultural production by the commercial nursery industry. Ornamental plants also include herbaceous and woody indoor and outdoor landscape broadleaf plants, grasses, and palms produced by traditional floricultural and nursery techniques within greenhouses, shaded structures, and other environments significantly modified to favor healthy, rapid, and profitable plant growth.

Ornamentals include annual, biennial, or perennial plants. They may be field grown in native or amended soils and then harvested and marketed with native soils intact. This form of horticulture is generally referred to as "balled and burlapped" (B&B) plant production even though burlap may not be used in their harvest. They may also be harvested without soil and referred to as "bare root." The most popular method of growing ornamental plants is in soilless growing media within containers. Soilless growing media are most often natural organic materials such as peat or tree bark mixed with a mineral component such as sand or perlite.

Ornamental plants comprise one of the economically and environmentally most important segments of American horticulture. U.S. Department of Agriculture farm income estimates from the production of greenhouse and nursery crops were $11 billion in 1997. California, Florida, Texas, and North Carolina were the top states producing ornamental plants.

Ornamental Plants for Garden

Like those beautiful flowers, ornamental plants can make your garden more enchanted. Ornamental plants are always improved each year. The plants that grow in the wild are planted as the collections for the garden.

Tamarind

This tree usually planted on the sidewalk. Tamarind has strong stem and generally used as bonsai also. This African tropic plant begins to be famous as a decoration plants. Tamarind can grow up to 50 feet. It has small brown acidic fruit. This tree can be planted beside your wall or pond.

Japanese Bamboo

Japanese Bamboo is categorized as Gramineae family. This plant is originated from Japan and China. They can grow up to 20 feet. This thick bamboo usually planted beside the wall. This bamboo can be used as natural fence. They can be planted between the rocks formation on your garden.

Dragon Tree

Dragon Tree is a sub tropic plants originated from the Canary Island. Dragon tree or Dracaena draco can grow up to 40 feet. It has thick leaves with the smooth surfaces colored in green. The Dragon Tree is the perfect choice to be planted in your garden corner. It can be used as the central plant by adding a variety of bedding plants.

Banyan Tree

This tree can be located in most of tropic nations. This tree can grow enormously huge up to 16 feet. Banyan leaves are colored in dark green and oval shaped. They can produce their roots that hanging in the air. Banyan tree can grow in a dry soil, good drainage and sufficient sunlight. Banyan tree usually planted to get an artistic effect and they can be a good decoration plant for Japanese style garden.

Rubber Banyan Tree

This banyan species has a big stem and it can grow taller. Rubber Banyan Tree or Ficus retuca originated from the subtropical area in the Southern part of China. This plant is suitable for Japanese style garden and usually planted as the main tree with the small plants accent around it.

Yellow Bamboo

This yellow colored bamboo is originated from China and Japan. They can grow up to 18 feet and the diameter is 8 cm. It has dark green leaves. Yellow Bamboo or Phyllostachys sulphurea has curvy leaves structures combined with golden yellow stem. This beautiful bamboo is suitable to be planted in group. Your fish pond will get extra natural scenery with this plant.

White Cedar

This tree shaped like a pyramid and they can grow up to 40 feet. White cedar's leaves are dark green and yellow on the tip of the leaves. The stem is reddish-brown. White cedar or Thuja occidentalis can be planted on the sideline along your fence. This plant will give a strong effect for giving directions before we enter the garden.

Mediterranean Cypress

Mediterranean Cypress or Cupressus sempervirens is a slender tress originated from South Europe, West Asia and North India. According to the history, these classic trees are always mentioned by the Greek and Romanian writers. The stems are very short and they can grow up to 40 feet. They can be planted beside the wall of your house or garden.

Australian Pine

This pine is popular as the bonsai structure. Australian pine grows a lot in the coastal area. The leaves like a spiny wire and this plant is suitable for your Japanese style garden.

Furcracea

This tropical plant is coming from Brazil. The leaves are shaped like a sword and slippery surfaces around 30-80 cm. The tip of the leaves is sharp like a thorn. Furcracea or Furcracea gigantean Striata is categorized into Amaryllidaceae family. This plant can be planted along the side of your wall and fence.

Juniperus

Another tree from Japan, the tree's stem is interesting with their spiral effect. Juniper's fruit like a small cherry. The fruit's colored in green and they turned into dark brown when it's ripe. This beautiful pine can be planted in line. You can also make this plant as your natural fence.

Stem Cactus

This giant cactus can grow for 20 feet. Stem cactus has black colored thorns. Stem cactus has yellow little flowers and poisonus sap. They can grow in a dry soil and high intensity of sunrise. Stem Cactus or Euphorbia canariensis and these plants should be planted beside your pond or near your fence.

Dates

This family member of Palmae can be located in dry sub tropic and tropic area with sufficient sunlight. Dates or Phoenix dactylifera is suitable as elemental plant in a wide garden. It can be planted in the corner of your garden or as the background of gradation plants. Dates will gives majestic and tough impression. The fruit from this tree can also be harvested.

Bactris Palm

This plant commonly found in Amazon and Costa Rica. The fruits shaped like hanging peach. Bactris palm or Bactris gasipaes can be planted in a wide area of your garden. It can also be planted in the tall side of your fence/wall.

Manila Palm

This tropical palm originated from Philippines and spread through another tropic areas. Manila palm or Veitchia merilli can grow up to 7 meter. The curved leaves are dark green. The fruits are colored in red which makes this plant more beautiful. Manila palm can grow rapidly and they are sensitive in cold weather and they can grow into maximum with a proper sunlight. Manila palm can be planted as a single plant or group with different height.

Triangular Palm

Triangular Palm or Neodypsis decaryi is coming from Madagascar. This palm species can be planted as a single tree in your left and right garden entrance to give friendly impression.

Dwarf Date Palm

Dwarf date or Phoenix robbelini usually called as the mini version of dates. This plant is originated from China and India. Their leaves are hung down with average length measured in 1,5 meters. The fruits are black. Their small appearance makes it easier for placing them as one of your garden elements. Dwarf date palm can be planted in line as the separator plants.

Ornamental Plants for Vertical Garden

In a vertical garden, we can grow a wide range of plants; some are edible, some strictly for beautifying purposes, while some are for both. Vertical gardens are a green solution to urban gardening problems within a limited space. Ornamental plants not only add to the aesthetic sense of the place but also help in making the space eco-friendly.

The various ornamental plants including both foliage and flowering plants.

Foliage

- Sword Fern: They are most common vertical garden plants used, nowadays. Often grown in hanging baskets on green walls.
- Botanical name: *Nephrolepis exaltata.*
- Light and temperature: Light to heavy shade; most ferns like an average temperature of 65-75 degrees F.
- Soil: pH should be around 7-8.
- Water: Regular watering and never let the soil get completely dry.
- Height: Up to 4-5 feet.
- Golden Pothos or Money plant: It is a vining plant that is super easy to care for. They easily grow and spread over a wide length of the wall. Botanical name: *Epipremnum aureum.*
- Light and temperature: Money plant prefers indirect light and will tolerate low light; thrives better in humid areas such as bathroom walls.
- Soil: Grows best in fertile loamy soil of pH range 6.0-7.5.
- Water: Needs watering every 7 to 10 days when the top 2 inches of soil is dry.
- Height: Grows up to a length of 20 feet.

- English ivy: Ivy is a hardy, evergreen vine. It attaches itself by aerial rootlets to the walls and other surfaces. Grown primarily for its lush foliage, which forms a cool green cover on walls or whichever vertical surface they're grown on.
- Botanical name: *Hedera helix.*

- Light and temperature: Full to partial sun; 50-70 degrees F.
- Soil: Well-drained loamy soil, either acidic, neutral or alkaline in nature.
- Water: Average watering is required.
- Height: 20-90 feet.

- Giant lilyturf: They have dense spreading strap-like leaves that are glossy and dark green. Though it is grown for its dense green foliage, it does bloom in summer and its flowers are quite showy.
- Botanical name: *Liriope gigantea.*
- Light and temperature: Full-sun to full-shade; 60-70 degrees F.
- Soil: well-drained loamy or clayey soil with pH ranging from neutral to acidic.
- Water: Needs regular watering throughout the growing season.
- Height: Up to 4 feet tall.
- Philodendron: It is easy to take care of, because if you watch the signals, the plant will tell you exactly what it needs. They are comfortable both as indoor and outdoor plants.
- Botanical name: *Philodendron scandens.*
- Light and temperature: Bright indirect sunlight; 75-85 degrees F.
- Soil: A soil of pH 6.0-6.5 is ideal for philodendron.
- Water: While watering philodendron plants, allow the top inch of the soil to dry out between watering.
- Height: Grows up to a length of 10-20 feet.

Flowers

- Geranium varieties: They bloom in burst of deep reds, scarlet, pinks, white, purple etc. with flowers lasting the entire gardening season.
- Botanical name: *Geranium dissectum.*
- Light and temperature: Bright to medium light is needed along with 60-70 degrees F.
- Soil: Well-drained potting soil of pH range 6.0-7.5.
- Water: Known to be drought tolerant, occasional watering can be carried out.
- Height: Grows up to 12 inches tall.
- Flower colours: Pink, blue, orange, white, red.

- Wedding vines: Wedding vine derives its name from the decoration purpose it serves at weddings. It's grown vertically widely for its pure white, waxy, intensely fragrant flowers.
- Botanical name: *Stephanotis floribunda.*
- Light and temperature: 70-90 degrees F.
- Soil: High content of loam and peat moss.
- Water: Water until the soil mixture is drenched; the mixture should drain freely.

- Height: Grows up to 6 feet or more.
- Flower colours: Pure white.
- Peace lily: Peace lily is used in a vertical garden owing to its quickly growing ability. Easy to grow as they are tolerant to lower light and low humidity.
- Botanical name: *Spathiphyllum wallisii.*
- Light and temperature: Bright to medium light; 65-80 degrees F temperature.
- Water: Water at least once a week and spritz the leaves with distilled water throughout the summer growing season.
- Height: Up to 2 feet (60 cm).
- Flowers: White.

- Star jasmine: Evergreen plant and can be easily trained on trellis.
- Botanical name: *Trachelospermum jasminoides.*
- Light and temperature: Full to partial sun.
- Water: Average watering is needed.
- Height: 3-9 metres.
- Flowers: White.

- English lavender: Evergreen garden herb thrives in summer heat and dry weather. They are also grown for its medicinal and culinary uses.
- Botanical name: *Lavandula.*
- Light and temperature: Full sun; 70-75 degrees F.
- Soil: Neutral to alkaline soil.
- Water: Regular watering is not required.
- Height: 1-2 metres.
- Flowers: Lavender in colour.

Plant Development

Plant development is an umbrella term for a broad spectrum of processes that include: the formation of a complete embryo from a zygote; seed germination; the elaboration of a mature vegetative plant from the embryo; the formation of flowers, fruits, and seeds; and many of the plant's responses to its environment. Plant development encompasses the growth and differentiation of cells, tissues, organs, and organ systems. Plant development shares many similarities with developmental processes in animals, but the fact that plants are nonmotile, photosynthetic organisms requires certain novel developmental processes in addition to the common ones.

Embryo and Seed Development

Embryogenesis, the formation of a multicellular embryo from a single-celled zygote, is one of the most dramatic and best-characterized aspects of plant development. Four key developmental processes take place during embryogenesis. First, the zygote expresses apical -basal polarity, meaning that the apical and basal ends of the zygote cell differ structurally and biochemically. When the zygote divides, it typically divides asymmetrically, giving rise to a small apical cell with dense cytoplasm and a large basal cell with watery cytoplasm. Although these two cells have identical nuclei, their fates differ dramatically. The apical cell gives rise to the embryo itself, while the basal cell gives rise to a short-lived structure called a suspensor and the tip of the root system. The progeny of the apical cell grow and divide to form a spherical mass of cells, the globular-stage embryo. Second, differential growth within the globular embryo gives rise to the "heart" stage embryo, the earliest stage when the precursors of cotyledons, root, and stem can be recognized. This key embryogenic process is called organogenesis. Third, distinctive planes of cell divisions bring about histogenesis, the process by which cells within embryonic cotyledons, root, and stem acquire different shapes, forming the precursors of the plant tissue systems. Last, the apical meristems of the shoot and root systems are formed at the apical and basal ends of the embryo.

After an embryo has reached full size, developmental changes continue to occur at the cellular level. Embryonic cells, particularly those of the cotyledons, begin to synthesize and store the proteins, lipids, and starch that will provide the energy and basic building blocks for germination and seedling growth. Next, the embryo begins to desiccate, sometimes losing up to 80 percent of its previous water content, and enters a phase of dormancy. Development and metabolism are arrested in dormant embryos, and seeds containing dormant embryos can survive for many years (sometimes centuries) and withstand extreme temperatures and drought.

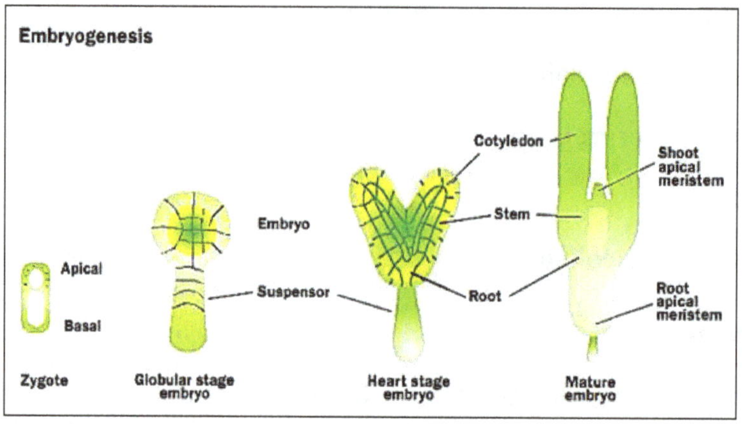

Embryo formation begins with cell division that establishes the apical-basal (top-bottom) axis. Further divisions elaborate on this basic plan, finally forming the cotyledons (seed leaves), as well as the apical meristems of root and shoot.

Plant hormones are important regulators of embryogenesis and seed dormancy. The hormones auxin, gibberellic acid, and cytokinin all stimulate growth and are present in the embryo during the stages of embryogenesis. As the embryo matures, these hormones are degraded and abscisic acid is synthesized by the embryo. Abscisic acid provides a developmental signal for the embryo to initiate the synthesis of storage compounds and to undergo desiccation. Abscisic acid is present in dormant seeds and is thought to play an important role in maintaining seed dormancy.

Germination and Seedling Development

Embryo development and metabolism resume upon seed germination. Given the right combination of water availability, temperatures, and light, the desiccated seed begins to take up water and the embryo begins to grow and metabolize again. Some species have specific requirements for germination; for instance, many temperate zone tree species require several weeks of temperatures of 4 degrees Celsius (39.2 degrees Fahrenheit) or less in order to germinate. Other species require low levels of light in order to germinate. Once germination is initiated, the embryo follows a typical pattern of development. In many plants, the preformed embryonic root elongates first, forcing its way out of the seed coat and into the soil. Next, the embryonic stem, usually the part below the attachment of the cotyledons (the hypocotyl), elongates. Once the hypocotyl has carried the cotyledons into the light, they expand, providing a broad surface for photosynthesis.

Environmental factors and their translation into hormonal signals are important for seedling development. For instance, germination in the dark results in developmental events that help the seedling push its way through the soil into the light. The hypocotyl elongates quickly and maintains a "hook" near its tip that protects the cotyledons and shoot apical meristem region. Cotyledon expansion is suppressed so that they are not damaged as they are pulled through the soil. In contrast, if the same seeds germinate in the light, the hypocotyl hardly elongates at all and does not form a hook, while the cotyledons quickly expand. The hormone gibberellic acid plays an important role in seed germination and early seedling growth. Gibberellic acid induces the synthesis of enzymes required for the metabolism of stored foods, thus providing energy for seedling growth. Gibberellic acid also induces cell division and cell expansion in dark-grown hypocotyls, maintaining their rapid growth through the soil.

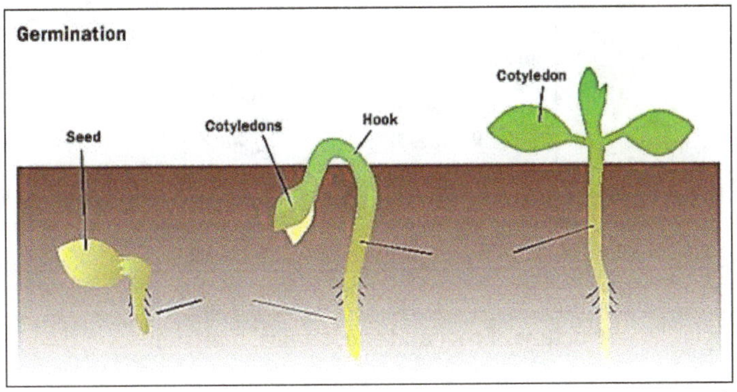

The root is the first portion of the plant to emerge during germination. Growth of the stem behind the cotyledons forms a "hook" that emerges from the soil, followed by emergence of the cotyledons, which begin to photosynthesize to feed further growth.

Apical Meristems and Development

The early stages of germination simply involve the enlargement of the root, hypocotyl, and cotyledons that were preformed in the embryo. Postembryonic development, however, is focused on the apical meristems. The shoot apical meristem is the source of all the leaves, stems, and their component cells formed during the lifetime of the plant. The meristem itself is composed of a small population of perpetually embryonic (meristematic) cells. These cells grow and divide, giving rise to new cells, but never mature themselves. Thus there is always a source of new cells at the tip of the shoot. The root tip has a similar population of meristematic cells that gives rise to all root tissues. Both of these meristems are characterized by an indeterminate growth pattern: one that is not finite, but, in theory at least, could continue throughout the lifetime of the plant.

Apical meristems are involved in several distinct developmental processes. The meristems are the location of cell proliferation and thus the source of all new cells in the shoot and root systems. The regions below the meristems are the sites of active growth, as new shoot and root tissue rapidly expands. The shoot apical meristem plays a role in organogenesis, the formation of new leaves and axillary buds in a precise spatial pattern. In contrast, the root apical meristem is not involved in organogenesis; lateral roots are initiated by pericycle cells, which are themselves derived from the meristem, usually several centimeters away from the meristem. The apical meristems also play a role in histogenesis by giving rise to cells that undergo distinct patterns of differentiation to form the specialized tissue types of the shoot and root. While the embryo initially gives rise to the precursors of dermal, ground, and vascular tissues (protoderm, ground meristem, and procambium, respectively), these tissue precursors continue to be formed by the apical meristems and represent the first stages of cell and tissue differentiation.

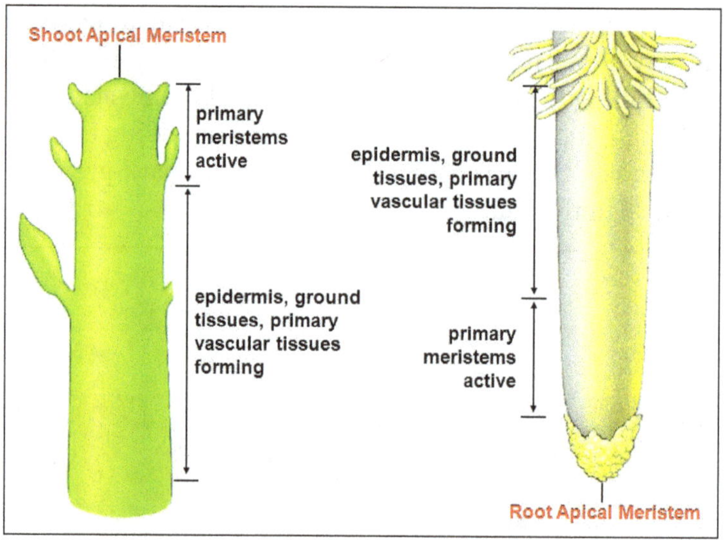

Structure of root and shoot apical meristems.

Cell Growth and Cell Division

Growth is defined as an irreversible increase in mass that is typically associated with an increase in volume. Plant cell growth is associated with meristems and must be carefully regulated in order for organogensis and histogenesis to occur in the appropriate patterns. The plant regulates growth by regulating theextensibility of its cell walls. A cell that has nonextensible cells walls can take up some water, but eventually the physical pressure of the water inside the cell pressing out on the cell wall (the turgor pressure) prevents the entry of additional water and any further change in volume. In contrast, a cell that has extensible cell walls can take up a substantial volume of water and thus increase in size. Turgor pressure that would otherwise prevent water entry momentarily decreases because the walls keep stretching.

Typically cell growth occurs in small increments: (1) wall extensibility increases, reducing turgor pressure; (2) reduced turgor pressure allows water to enter the cell, increasing cell volume; (3) wall extensibility decreases, allowing the cell to build up turgor and preventing further water entry; and (4) the cell undergoes a cycle of synthesis of cytoplasmic and wall components, adding to the cell's mass. This cycle of incremental growth is repeated many times until the cell reaches its final size.

The plant hormones auxin and gibberellin are produced in the vicinity of the apical meristems and usually act in concert to induce cell growth. Both hormones regulate wall extensibility, but carry out this function in different ways. Auxin induces the activity of cell membrane H + adenosine triphosphatase (ATPase) molecules. Proton (H +) extrusion lowers the pH of the cell wall, thus activating the cell wall enzyme expansin. Expansin cleaves the hydrogen bonds between two cell wall components: The cellulose microfibrils and the hemicellulose molecules that link adjacent cellulose microfibrils. Breakage of these bonds allows these structural wall components to reposition themselves farther apart, increasing wall extensibility. Gibberellic acid, on the other hand, stimulates the activity of another cell wall enzyme called xyloglucan endotransglycosylase (XET). Xyloglucans are a type of hemicellulose that is cleaved by the XET enzyme. Breakage of the hemicellulose molecules also allows the cellulose microfibrils to move farther apart, increasing wall extensibility.

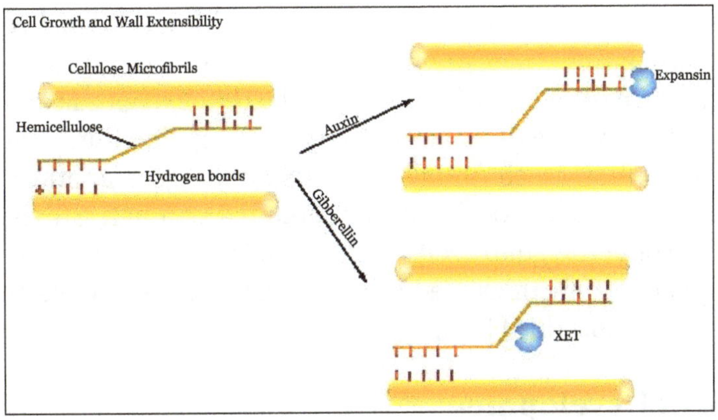

The hormones auxin and gibberellin each promote cell expansion by loosening the bonds between adjacent cell wall molecules. Each hormone acts on a different molecular target.

Cell division and cell growth are often tightly linked. When the rate of cell division is balanced by cell growth, as in the apical meristems, average cell size does not increase. As the meristem grows away from earlier formed cells, the ratio of growth to division increases, resulting in overall cell enlargement. As the tissues mature further, cell division ceases completely, giving rise to zones of pure cell enlargement where most of the visible growth of the plant occurs. This relationship between division and growth, coupled with observations of the predictable planes of cell division during histogenesis, indicates that cell division is carefully regulated during plant development.

Molecules called cyclin-dependent kinases (CDKs) are key regulators of cell cycling (including cell division) in plants. CDKs are activated by association with a regulatory subunit called a cyclin and by phosphorylation and dephosphorylation events. The plant hormone cytokinin appears to regulate the cell cycle by interacting with the CDKs. Cytokinins enhance the synthesis of the cyclin subunits that are required for the cell to enter the deoxyribonucleic acid (DNA) synthesis phase of the cell cycle. Cytokinins also enhance the CDK dephosphorylation step that is required for the cell to progress into mitosis. Both of these processes are inhibited by the hormone abscisic acid; thus a "developmental tug-of-war" occurs between a division-enhancing hormone and a division-suppressing hormone. The delicate balance between them determines the rate of cell division and this type of interaction is probably typical of the hormonal regulation of many aspects of plant development.

Differentiation

Differentiation is the process whereby cells, tissues, and organs become different from each other and from their precursors. The concept can be applied to organogenesis since cotyledons, foliage leaves, sepals, and petals may all develop from similar appearing precursors, the leaf primordia. As these organs mature, they become different from each other in size, shape, and the development of distinctive cell types. For instance, the epidermis tissue of petals is sharply differentiated from that of cotyledons, foliage leaves and sepals that are photosynthetic organs. Correlated with a photosynthetic function, the epidermis of these organs is made up of flat, transparent cells that allow the penetration of light into internal tissues. Specialized guard cells that allow CO_2 to enter the leaf are also present. In contrast, the epidermal cells of petals contain brightly colored carotenoid or anthocyanin pigments. These cells also have a papillate shape that imparts a velvetlike sheen to the petal surface. Since petals carry on minimal photosynthesis, they often lack guard cells.

The process of differentiation is best understood on a cellular level. For instance, guard cells are highly specialized epidermal cells. Early in the development of a leaf, protodermal precursor cells undergo a distinctive pattern of cell divisions. At first the cell divisions are asymmetric, producing one large and one small derivative. The large derivative stops dividing and differentiates as an unspecialized epidermal cell, while the small derivative undergoes another asymmetric division. At an unknown stop signal, the small derivative undergoes a symmetric division, giving rise to two equal sized cells that become the guard cells. Unlike their plain neighbors, these cells develop a distinctive kidney shape, unevenly thickened cell walls, large, conspicuous chloroplasts, and finally form a pore (the stomatal aperture) between them.

Uniqueness of Plant Development

Although plants share many features of development with animals such as apical-basal polarity, regulation of the balance between cell growth and cell division, formation of distinctive patterns of organs, cells and tissues, and differentiation, some aspects of development are unique to plants. Among these are:

- The formation and maintenance of the perpetually embryonic regions, the apical meristems. The meristems have an indeterminate growth pattern that result in the occurrence of growth, organogenesis, and histogenesis throughout the life of the plant.
- Plant cells have rigid cell walls that prevent cell movement. Thus organogenesis and histogenesis must occur through differential growth and regulation of the planes of cell division. Cell-cell communication is important in plant development, but cell recognition is likely less important than it is in animals since plant cells keep the same neighbors throughout their life.
- Plant cells are totipotent; that is, able to differentiate as a different cell type if given the appropriate stimulus. Totipotency is likely a reflection of the plant's sedentary lifestyle. Plants can't escape predators and other kinds of damage, but they can readily repair wounds and reconnect vascular strands by differentiating the appropriate cell types.

Plant Growth

Most plants continue to grow throughout their lives. Like other multicellular organisms, plants grow through a combination of cell growth and cell division. Cell growth increases cell size, while cell division (mitosis) increases the number of cells. As plant cells grow, they also become specialized into different cell types through cellular differentiation. Once cells differentiate, they can no longer divide. How do plants grow or replace damaged cells after that?

The key to continued growth and repair of plant cells is meristem. Meristem is a type of plant tissue consisting of undifferentiated cells that can continue to divide and differentiate.

Apical meristems are found at the apex, or tip, of roots and buds, allowing roots and stems to grow in length and leaves and flowers to differentiate. Roots and stems grow in length because the meristem adds tissue "behind" it, constantly propelling itself further into the ground (for roots) or air (for stems). Often, the apical meristem of a single branch will become dominant, suppressing the growth of meristems on other branches and leading to the development of a single trunk. In grasses, meristems at the base of the leaf blades allow for regrowth after grazing by herbivores—or mowing by lawnmowers.

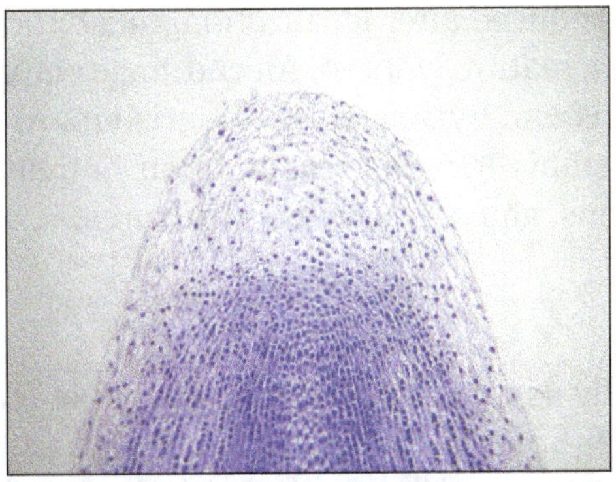

Microphotograph of the root tip of a broad bean show rapidly dividing apical vmeristem tissue just behind the root cap. Numerous cells in various stages of mitosis can be observed.

Apical meristems differentiate into the three basic types of meristem tissue which correspond to the three types of tissue: protoderm produces new epidermis, ground meristem produces ground tissue, and procambium produces new xylem and phloem. These three types of meristem are considered primary meristem because they allow growth in length or height, which is known as primary growth.

Secondary meristems allow growth in diameter (secondary growth) in woody plants. Herbaceous plants do not have secondary growth. The two types of secondary meristem are both named cambium, meaning "exchange" or "change." Vascular cambium produces secondary xylem (toward the center of the stem or root) and phloem (toward the outside of the stem or root), adding growth to the diameter of the plant. This process produces wood, and builds the sturdy trunks of trees. Cork cambium lies between the epidermis and the phloem, and replaces the epidermis of roots and stems with bark, one layer of which is cork.

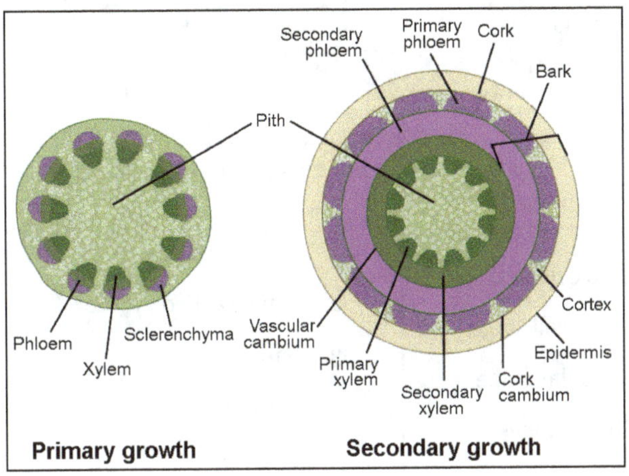

Woody plants grow in two ways: Primary growth adds length or height, mediated by apical meristem tissue at the tips of roots and shoots—which is difficult to show clearly in cross-sectional figure. Secondary growth adds to the diameter of a stem or root; vascular cambium adds xylem (inward) and phloem (outward), and cork cambium replaces epidermis with bark.

Plant Embryogenesis

Plant embryogenesis is a process that occurs after the fertilization of an ovule to produce a fully developed plant embryo. This is a pertinent stage in the plant life cycle that is followed by dormancy and germination. The zygote produced after fertilization, must undergo various cellular divisions and differentiations to become a mature embryo. An end stage embryo has five major components including the shoot apical meristem, hypocotyl, root meristem, root cap, and cotyledons. Unlike animal embryogenesis, plant embryogenesis results in an immature form of the plant, lacking most structures like leaves, stems, and reproductive structures.

Morphogenic Events

Embryogenesis occurs naturally as a result of single, or double fertilzation, of the ovule, giving rise to two distinct structures: the plant embryo and the endosperm which go on to develop into a seed. The zygote goes through various cellular differentiations and divisions in order to produce a mature embryo. These morphogenic events form the basic cellular pattern for the development of the shoot-root body and the primary tissue layers; it also programs the regions of meristematic tissue formation. The following morphogenic events are only particular to eudicots, and not monocots.

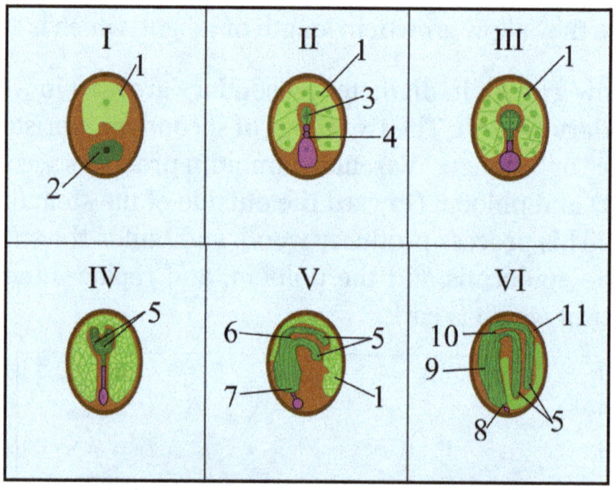

This depicts six moments in embryogenesis.

I.	Two cell stage	1. endosperm
II.	Eight cell stage	2. single celled zygote
III.	Globular stage	3. embryo
IV.	Heart stage	4. suspensor
V.	Torpedo stage	5. cotyledons
VI.	Maturation	6. shoot apical meristem (SAM)
		7. root apical meristem (RAM)

Two Cell Stage

Following fertilization, the zygote and endosperm are present within the ovule, as seen in stage I of the illustration on this page. Then the zygote undergoes an asymmetric transverse cell division that gives rise to two cells - a small apical cell resting above a large basal cell. These two cells are very different, and give rise to different structures, establishing polarity in the embryo.

Apical Cell

The small apical cell is on the top and contains most of the cytoplasm, the aqueous substance found within cells, from the original zygote. It gives rise to the hypocotyl, shoot apical meristem, and cotyledons.

Basal Cell

The large basal cell is on the bottom and consists of a large vacuole and gives rise to the hypophysis and the suspensor.

Eight Cell Stage

After two rounds of longitudinal division, and one round of transverse division, an eight-celled embryo is the result. Stage II, in the illustration above, indicates what the embryo looks like during the eight cell stage. According to Laux et al., there are four distinct domains during the eight cell stage. The first two domains contribute to the embryo proper. The apical embryo domain, gives rise to the shoot apical meristem and cotyledons. The second domain, the central embryo domain, gives rise to the hypocotyl, root apical meristem, and parts of the cotyledons. The third domain, the basal embryo domain, contains the hypophysis. The hypophysis will later give rise to the radicle and the root cap. The last domain, the suspensor, is the region at the very bottom, which connects the embryo to the endosperm for nutritional purposes.

Sixteen Cell Stage

Additional cell divisions occur, which leads to the sixteen cell stage. The four domains are still present, but they are more defined with the presence of more cells. The important aspect of this stage is the introduction of the protoderm, which is meristematic tissue that will give rise to the epidermis. The protoderm is the outermost layer of cells in the embryo proper.

Globular Stage

The name of this stage is indicative of the embryo's appearance at this point in embryogenesis; it is spherical or globular. Stage III, in the photograph above, depicts what the embryo looks like during the globular stage. 1 is indicating the location of the endosperm. The important component of the globular phase is the introduction of the rest of the primary meristematic tissue. The protoderm was already introduced during the sixteen cell stage. According to Evert and Eichhorn, the ground meristem and procambium are initiated during the globular stage. The ground meristem will go on to form the ground tissue, which includes the pith and cortex. The procambium will eventually form the vascular tissue, which includes the xylem and phloem.

Heart Stage

According to Evert and Eichhorn, the heart stage is a transition period where the cotyledons finally start to form and elongate. It is given this name in eudicots because most plants from this group have two cotyledons, giving the embryo a heart shaped appearance. The shoot apical meristem is between the cotyledons. Stage IV, in the illustration above, indicates what the embryo looks like at this point in development. 5 indicates the position of the cotyledons.

Torpedo Stage

This stage is defined by the continued growth of the cotyledons and axis elongation. In addition, programmed cell death must occur during this stage. This is carried out throughout the entire growth process, like any other development. However, in the torpedo stage of development, parts of the suspensor complex must be terminated. The suspensor complex is shortened because at this point in development most of the nutrition from the endosperm has been utilized, and there must be space for the mature embryo. After the suspensor complex is gone, the embryo is fully developed. Stage V, in the illustration above, indicates what the embryo looks like at this point in development.

Maturation

The second phase, or postembryonic development, involves the maturation of cells, which involves cell growth and the storage of macromolecules (such as oils, starches and proteins) required as a 'food and energy supply' during germination and seedling growth. The appearance of a mature embryo is seen in Stage VI.

Dormancy

The end of embryogenesis is defined by an arrested development phase, or stop in growth. This phase usually coincides with a necessary component of growth called dormancy. Dormancy is a period in which a seed cannot germinate, even under optimal environmental conditions, until a specific requirement is met. Breaking dormancy, or finding the specific requirement of the seed, can be rather difficult. For example, a seed coat can be extremely thick. According to Evert and Eichhorn, very thick seed coats must undergo a process called scarification, in order to deteriorate the coating. In other cases, seeds must experience stratification. This process exposes the seed to certain environmental conditions, like cold or smoke, to break dormancy and initiate germination.

Role of Auxin

Auxin is a hormone related to the elongation and regulation of plants. It also plays an important role in the establishment polarity with the plant embryo. Research, conducted by Cooke, Racusen, and Cohen, has shown that the hypocotyl from both gymnosperms and angiosperms show auxin transport to the root end of the embryo They hypothesized that the embryonic pattern is regulated by the auxin transport mechanism, and the polar positioning of cells within the ovule. The importance of auxin was shown, in their research, when carrot embryos, at different stages, were subjected to auxin transport inhibitors. The inhibitors that these carrots were subjected to made them unable to progress to later stages of embryogenesis. During the globular stage of embryogenesis, the embryos continued spherical expansion. In addition, oblong embryos continued axial growth, without the introduction of cotyledons. During the heart embryo stage of development there were additional growth axes on hypocotyls. Further auxin transport inhibition research, conducted on Brassica juncea, show that after germination, the cotyledons were fused and not two separate structures.

Alternative Forms of Embryogenesis

Somatic Embryogenesis

Somatic embryos are formed from plant cells that are not normally involved in the development of embryos, i.e. ordinary plant tissue. No endosperm or seed coat is formed around a somatic embryo. Applications of this process include: clonal propagation of genetically uniform plant material; elimination of viruses; provision of source tissue for genetic transformation; generation of whole plants from single cells called protoplasts; development of synthetic seed technology. Cells derived from competent source tissue are cultured to form an undifferentiated mass of cells called a callus. Plant growth regulators in the tissue culture medium can be manipulated to induce callus formation and subsequently changed to induce embryos to form the callus. The ratio of different plant growth regulators required to induce callus or embryo formation varies with the type of plant. Asymmetrical cell division also seems to be important in the development of somatic embryos, and while failure to form the suspensor cell is lethal to zygotic embryos, it is not lethal for somatic embryos.

Androgenesis

The process of androgenesis allows a mature plant embryo to form from a reduced, or immature, pollen grain. Androgenesis usually occurs under stressful conditions. Embryos that result from this mechanism can germinate into fully functional plants. As mentioned, the embryo results from a single pollen grain. Pollen grains consists of three cells - one vegetative cell containg two generative cells. According to Maraschin et al., androgenesis must be triggered during the asymmetric division of microspores. However, once the vegetative cell starts to make starch and proteins, androgenesis can no longer occur. Maraschin et al., indicates that this mode of embryogenesis consists of three phases. The first phase is the *acquisition of embryonic potential*, which is the repression of gametophyte formation, so that the differentiation of cells can occur. Then during the *initiation of cell divisions*, multicellular structures begin to form, which are contained by the exine wall. The last step of androgenesis is *pattern formation*, where the embryo-like structures are released out of the exile wall, in order for pattern formation to continue.

After these three phases occur, the rest of the process falls in line with the standard embryogenesis events.

Plant Growth and Buds

Embryonic tissue is made up of actively growing cells and the term is normally used to describe the early formation of tissue in the first stages of growth. It can refer to different stages of the sporophyte and gametophyte plant; including the growth of embryos in seedlings, and to meristematic tissues, which are in a persistently embryonic state, to the growth of new buds on stems.

In both gymnosperms and angiosperms, the young plant contained in the seed, begins as a developing egg-cell formed after fertilization (sometimes without fertilization in a process called apomixis) and becomes a plant embryo. This embryonic condition also occurs in the buds that form on stems. The buds have tissue that has differentiated but not grown into complete structures. They can be in a resting state, lying dormant over winter or when conditions are dry, and then commence growth when conditions become suitable. Before they start growing into stem, leaves, or flowers, the buds are said to be in an embryonic state.

Plug

Plugs in horticulture are small-sized seedlings grown in trays from expanded polystyrene or polythene filled usually with a peat or compost substrate. This type of plug is used for commercially raising vegetables and bedding plants. Similarly plugs may also refer to small sections of lawn grass sod. After being planted, lawn grass may somewhat spread over an adjacent area.

Plug plants are young plants raised in small, individual cells, ready to be transplanted into containers or a garden. Professionally raised vegetable/flowering plants in controlled conditions during their important formative period (the first 4–6 weeks) can help to ensure plant health and for plants to reach their maximum potential during the harvest/blooming period. Establishing a garden using plug plants is often easier than doing so starting from seed.

Plugs growing in a greenhouse at New Place Nurseries in Pulborough.

Planting from plugs reduces the time a crop resides in the ground, and is functional for those with limited space. Plugs can improve yields: a healthy, stocky plant will grow rapidly and symmetrically when planted out, with a potentially greater capacity to withstand pests, disease and drought. Raising some types of seedlings successfully can be difficult, so plug plants can be beneficial for less experienced gardeners. Plug plants are beneficial for gardeners who want to try a new variety or a range of varieties without purchasing numerous packets of seeds and starting the plants from seed. Plug plants are very useful if the sowing window is missed, and plugs can be purchased quickly to replace a crop which has failed.

As a garden develops, interplanting (intercropping) existing crops with plugs plants, ideally companion plants, can improve the productivity of the space and so maximise harvests – a sown crop may not be able to compete with established plants. Plug plants are much easier to weed than sown seedlings, and weeding will need to be done less frequently.

Having semi-grown plants simplifies designing a vegetable plot or container. As plants that have already started growth, the time to attain plant growth is lessened. Within days of planting signs of growth are typically visible: leaves will perk up and roots anchor into the soil. Air pruned plugs are grown in a manner to promote very rapid growth almost immediately after being transplanted to new soil.

Plugs are sometimes used in hillside plasticulture applications, due to the ease in which they are transplanted.

Plant Cultivation and Growth

A plug of St. Augustine grass ready for sprigging.

Plug plants grow more consistently, as has been noted by the commercial scale vegetable growing industry, and more rapidly; large-scale brassica field crops are planted almost exclusively from soil block plugs in some parts of Europe, a trend which is growing in the UK. This success at the commercial scale is testament to the success of plugs in the ground.

It is of note that many varieties actively benefit from being transplanted as severing the taproot encourages bushier root growth. Traditionally nearly all heading brassica are sown in a separate seed bed, thinned, and the best ones planted in a prepared bed after about 6–8 weeks. Many pests want to eat baby brassica; this in combination with its long growing season makes planting brassica from plugs a much easier option.

Root Vegetables

Root vegetables are typically, but not always, sown from seed, rather than transplanted from plugs, where they are to mature and then be thinned. The thinning action is highly beneficial in itself as it provides soil aeration at depth without disturbing adjacent roots systems. The initial concentration of seedlings also dilutes damage from pests and provided some food for the gardener or the compost in the form of thinnings. Beetroot, carrots and the root brassica family- swede, turnip- will simply not reach their full potential with any check to early root growth. In addition, these seeds are typically inexpensive, and the seedlings are delicate; hence there is little value to the gardener in buying or growing them as plugs.

Bedding

Bedding, in horticulture, refers to the temporary planting of fast-growing plants into flower beds to create colourful, temporary, seasonal displays, during spring, summer or winter. Plants used for bedding are generally annuals, biennials or tender perennials; succulents are gaining in popularity.

Some bedding plants are also referred to as "patio plants" because they are widely used in pots and other containers positioned on patios, terraces, decking and other areas around houses. Larger tender "conservatory plants" may also be moved out from greenhouses or conservatories and planted out in borders (or stood in their pots in sheltered positions) for the warmer months, then returned to shelter for the winter.

The modern bedding plant industry breeds and produces plants with a neat, dwarf habit, which flower uniformly and reliably. They are bred primarily for use in large-scale bedding schemes where uniformity and predictability is of paramount importance, but this is often achieved by losing the plants' individual character, and has been criticised by such notable plantsmen as the late Christopher Lloyd, who championed an informal style of bedding.

Bedding Plants

There exists a huge range of plants specifically grown to produce a period of flower colour throughout the spring and summer, and (usually) discarded after flowering. They may conveniently be divided into four groups:

- Hardy annuals sown directly into the ground early in the season (poppy, stock, sunflower, clarkia, godetia, eschscholzia, nigella, dianthus).
- Tender annual or perennial plants treated as half-hardy annuals – sown under glass in late winter in heat, or purchased as young plants, and hardened-off outdoors when all danger of frost has passed (begonia, lobelia, petunia, argyranthemum, chrysanthemum, pelargonium, nicotiana, cosmos, fuchsia).
- Hardy biennial plants, or perennials treated as biennial, sown in one year to flower the next, and discarded after flowering (antirrhinum, polyanthus, wallflower, daisy, foxglove, some dianthus, some poppies, campanula, delphinium, aubrieta, aquilegia, cornflower, pansies).
- Corms, rhizomes, bulbs and tubers, planted each year and lifted after the plant has died down and stored in winter, or discarded (tulip, narcissus, hyacinth, gladiolus, dahlia, canna).

Types of Bedding

Formal bedding, as seen in parks and large gardens, where whole flower beds are replanted two or three times a year, is a costly and labour-intensive process. Towns and cities are encouraged to produce impressive displays by campaigns such as "Britain in bloom".

Spring bedding (tulips and polyanthus), South Shields.

Spring Bedding

Plants used for spring bedding are often biennials (sown one year to flower the next), or hardy, but short-lived, perennials. Spring-flowering bulbs such as tulips are often used, typically with forget-me-nots, wallflowers, winter pansies and polyanthus.

Summer Bedding

Plants used for summer bedding are generally annuals or tender perennials. They become available (often as what are referred to as "plug plants") in nurseries and garden centres during spring, to be gradually "hardened off" (acclimatised to outdoor conditions) by the purchaser and finally planted out around the time that the last frosts are expected. Experienced gardeners keep an eye on the weather forecasts at that time of year and are on standby to protect their bedding displays overnight with horticultural fleece (or the older alternatives of net curtains or newspaper) if frost threatens.

Formal summer bedding scheme in Victoria Park, Bath, Somerset.
Most of the colour is provided by large-flowered begonias.

Carpet Bedding

Carpet bedding employs two or more contrasting plant cultivars with a neat, dwarf habit and distinct colouring (of flower or foliage) to create geometric displays. It is often used to form such things as lettering, logos or trademarks, coats of arms, or floral clocks. Suitable plants are rosette-forming succulents such as Echeveria or fairly slow-growing or mat-forming foliage plants, such as coloured-leaved Alternanthera cultivars, which are tolerant of clipping; such plants may also be used in three-dimensional sculptural forms or pseudo-topiary.

Municipal carpet bedding schemes often incorporate lettering.

Winter Bedding

Planted in autumn to give a display until early spring, the plants used for winter bedding are mainly hardy perennials. As it has to be planted at the same time of year as spring bedding does, winter bedding tends to be less commonly seen, except in containers such as windowboxes. Some are short-lived and will be discarded after their first display; others may be used as a source of cuttings for the next year. Winter-hardy ornamental vegetables such as cultivars of kale and cabbage with coloured or variegated foliage are increasingly common. Primula cultivars (polyanthus and primroses) are commonly used, as are winter-flowering heathers and Viola ×wittrockiana, winter pansies. Variegated evergreens such as cultivars of Vinca minor (lesser periwinkle), Euonymus fortunei and Hedera helix (ivies) are also popular.

Bokashi

Bokashi is a process that converts food waste and similar organic matter into a soil amendment which adds nutrients and improves soil texture. It differs from traditional composting methods in several respects. The most important are:

- The input matter is fermented by specialist bacteria, not decomposed.
- The fermented matter is fed directly to field or garden soil, without requiring further time to mature.
- As a result virtually all input carbon, energy and nutrients enter the soil food web, having been neither emitted in greenhouse gases and heat nor leached out.

Other names attributed to this process include bokashi composting, bokashi fermentation and fermented composting.

Process

The basic stages of the process are:

- Organic matter is inoculated with *Lactobacilli*. These will convert a fraction of the carbohydrates in the input to lactic acid by homolactic fermentation.

A household bokashi bin with a supply of fermentation starter, namely bran inoculated with Lactobacilli.

- Fermented anaerobically for a few weeks at typical room temperatures in an airtight vessel, the organic matter is altered and preserved, in a process closely related to the making of some fermented foods and silage. The preserve is normally applied to soil when ready, or can be stored unopened for later use.
- The preserve is mixed into soil. This exposes it to air, whereupon the lactic acid oxidises to pyruvate, a fundamental energy carrier in biological processes.
- The oxidised preserve is soon consumed by the indigenous soil life, 'disappearing' within a very few weeks at normal temperatures. Earthworm action is typically prominent, such that the amended soil acquires a texture associated with vermicompost.

Characteristics

Accepted Inputs

Inside a recently started bokashi bin. Food scraps are raised on a perforated plate (to drain runoff) and are partly covered by a layer of bran.

The process is typically applied to food waste from households, workplaces and catering establishments, because such waste normally holds a good proportion of carbohydrates. It is applied to other organic waste by supplementing carbohydrates and hence lactic acid production. Recipes for large scale bokashi in horticulture often include rice, and molasses or sugar. Any carbohydrate-poor waste stream would benefit from this.

Homolactic fermentation can process significantly more kinds of food waste than home composting. Even items such as cooked leftovers, meat and skin, fat, cheese and citrus waste are, in effect, pre-digested to enable soil life to consume them.

Pieces of input are discarded if they are already badly rotten, or show green or black mould. These harbour putrefying organisms which may overwhelm the fermentation.

Large items, and shapes with holes or depressions, can create air pockets which degrade the anaerobic environment. However most such items can be cut, crushed or broken up, so in practice this excludes only strong bones or similar.

Emissions

Carbon, Gases and Energy

Homolactic fermentation breaks no carbon bonds and emits no gas; its overall equation is $C_6H_{12}O_6$ (carbohydrate) → 2 $CH_3CHOHCOOH$ (lactic acid). It is a mildly endothermic reaction, emitting no energy; the fermentation vessel remains at ambient temperature.

These are in marked contrast to decomposition, which emits the majority of its input carbon and energy within greenhouse gases (carbon dioxide and methane) and - in aerobic decomposition - as heat. In addition, decomposition emits nitrogen - a key nutrient - within ammonia and the potent greenhouse gas nitrous oxide.

Runoff

When fermentation begins, physical structures start to break down and release some of the input's water content as a liquid runoff. Over time this constitutes more than 10% of the input by weight. The quantity varies with the input: for example cucumber and melon flesh lead to a noticeable increase.

The liquid leaches out a valuable fraction of proteins, nutrients and lactic acid. To recover them, and to avoid drowning the fermentation, runoff is captured from the fermentation vessel, either through a tap, into a base of absorbent material such as biochar or waste cardboard, or into a lower chamber. The runoff is sometimes called "bokashi tea".

The uses of bokashi tea are not the same as those of "compost tea". It is used most effectively when diluted and aerated (thus oxidising lactic acid to pyruvate) and then immediately sprinkled over a targeted area of soil. This ensures that all input food value is retained to feed the soil ecosystem, along with the later solid output. Myths have encouraged less effective uses that are either potentially damaging (e.g. feeding plants with acidic water) or wasteful (e.g. cleaning drains with nutrients, feeding plants with proteins they cannot take up).

Volumes

Household containers ("bokashi bins") typically give a batch size of 5-10 kilograms. This is accumulated over a few weeks of regular additions. Each regular addition is best accumulated in a caddy, because it is recommended that one opens the bokashi bin no more frequently than once per day to let anaerobic conditions predominate.

In horticultural settings batches can be orders of magnitude greater. Silage technology may be usable if it is adapted to capture runoff. An industrial-scale technique mimics the windrows of large-scale composting, except that bokashi windrows are compacted, covered tightly and left undisturbed, all to promote anaerobic conditions. One study suggests that such windrows lose only minor amounts of carbon, energy and nitrogen.

Hygiene

Lactic acid is a strong natural bactericide, with well-known antimicrobial properties. It is an active ingredient of some toilet cleaners. As more is produced, it eventually suppresses even its own makers, the acid-resistant lactobacilli, such that bokashi fermentation eventually slows and stops itself. There is also evidence that mesophilic (ambient temperature) fermentation kills eggs of the Ascaris worm - a parasite of humans - in 14 days.

The fermentation bin does not release smells when it is closed. A household bin is only opened for a minute or so to add and inoculate input via the lid or to drain runoff via the tap. At these times the user encounters the sour odour of lacto-fermentation (often described as a "pickle") which is much less offensive than the odour of decomposition.

An airtight fermentation bin cannot attract insects. Bokashi literature claims that scavengers dislike the fermented matter and avoid it in gardens.

Addition to Soil

Fermented bokashi is added to a suitable area of soil. The approach usually recommended by suppliers of household bokashi is along the lines of "Dig a trench in the soil in your garden, add the waste and cover over."

In practice, regularly finding suitable sites for trenches that will later underlie plants is difficult in an established plot. To address this, an alternative is a 'soil factory'. This is a bounded area of soil into which several loads of bokashi preserve are mixed over time. Amended soil can be taken from it for use elsewhere. It may be of any size. It may be permanently sited or in rotation. It may be enclosed, wire-netted or covered to keep out surface animals. Spent soil or compost, and organic amendments such as biochar may be added, as may non-fermented material, in which case the boundary between bokashi and composting becomes blurred.

A relatively new alternative is to homogenise (and potentially dilute) the preserve into a slurry, which is spread on the soil surface and mixed in by raking. The approach requires energy for homogenisation but confers several advantages. It thoroughly oxidises the preserve. It disturbs no deeper layers, except by increased worm action. Once mixed in, the slurry is of little use to scavenging animals. A larger area can be amended, and done repeatedly to sustain a larger soil ecosystem.

The approach should be scalable to professional volumes based on conventional spreading equipment.

Uses

The primary use of bokashi is to recover value from organic waste by converting it into a soil amendment. In Europe, food and drink material that is sent to animal feed does not legally constitute waste because it is regarded as 'redistribution'. This may apply to bokashi made from food, because it enters the soil food web, and furthermore is inherently pathogen-free.

A side effect of diverting organic waste to the soil food web is to divert it away from local waste management streams and their associated costs of collection and disposal. To encourage this, for example most UK local authorities subsidise household bokashi starter kits through a National Home Composting Framework. Another side effect is to increase the organic carbon content of the amended soil. Some of this is a relatively long-term carbon sink – insofar as the soil ecosystem creates humus – and some is temporary for as long as the richer ecosystem is sustained by measures such as permanent planting, no-till cultivation and organic mulch. An example of these measures is seen at the Ferme du Bec Hellouin in France. Bokashi would therefore have potential uses in enabling communities to speed up the conversion of land from chemical to organic horticulture and agriculture, to regenerate degraded soil, and to develop urban and peri-urban horticulture close to the sources of input.

The anti-pathogenic nature of bokashi is applied to sanitation, in particular to the treatment of faeces. Equipment and supplies to treat pet faeces are sold commercially but do not always give prominence to the hygiene risks.Treatment of human faeces has been extensively studied, notably with the use of biochar (a soil improver in its own right) to remove odours and retain nutrients. Social acceptability is a major obstacle, but niche markets such as emergency aid sanitation, outdoor events and temporary workplaces may develop the technology into a disruptive innovation. The anti-pathogenic nature of bokashi is the modern treatment of water pollution caused by faecal sludge, as the bokashi nature ferment the sludge & take away bad odor of the sludge and the water itself.

Bolting

While plants do not "run away" physically, their growth may run away rapidly, and this is basically what this phrase means in the gardening world. Plants, mostly vegetable or herbs, are said to bolt when their growth goes rapidly from being mostly leaf based to being mostly flower and seed based.

Why do Plants Bolt?

Most plants bolt due to hot weather. When the ground temperature goes above a certain temperature, this flips a switch in the plant to produce flowers and seeds very rapidly and to abandon leaf growth almost completely. Bolting is a survival mechanism in a plant. If the weather get to be above where the plant will survive, it will try to produce the next generation (seeds) as quickly as possible. Some plants that are known for bolting are broccoli, cilantro, basil, cabbage and lettuce.

Can you Eat a Plant after it Bolts?

Once a plant has fully bolted, the plant is normally inedible. The plant's entire energy reserve is focused on producing the seeds, so the rest of the plant tends to become tough and woody as well as tasteless or even bitter. Occasionally, if you catch a plant in the very early stages of bolting, you can temporarily reverse the process of bolting by snipping off the flowers and flower buds. In some plants, like basil, the plant will resume producing leaves and will stop bolting. In many plants though, such as broccoli and lettuce, this step only allows you some extra time to harvest the crop before it becomes inedible.

Preventing Bolting

Bolting can be prevented by either planting early in the spring so that bolt-prone plants grow during late spring, or late in the summer so they grow during early fall. You can also add mulch and ground cover to the area, as well as watering regularly in order to keep the soil temperature down.

References

- Ornamental-plants, news-wires-white-papers-and-books, science: encyclopedia.com, Retrieved 25 February, 2019
- 17-ornamental-plants-for-your-garden: invorma.com, Retrieved 19 April, 2019
- Maraschin, s. F.; de priester, w.; spaink, h. P.; wang, m. (july 2005). "androgenic switch: an example of plant embryogenesis from the male gametophyte perspective". Journal of experimental botany. 56 (417): 1711–1726. Doi:10.1093/jxb/eri190. Pmid 15928015
- Top-10-ornamental-plants-vertical-garden: greenmylife.in, Retrieved 17 May, 2019
- Dunnett, nigel; et al. (2011). Small green roofs: low-tech options for greener living. Timber press. Pp. 67–68. Retrieved march 30, 2012. Isbn 9781604690590
- Plant-development: biologyreference.com, Retrieved 14 July, 2019
- Souter, martin; lindsey, keith (june 2000). "polarity and signaling in plant embryogenesis". Journal of experimental botany. 51 (347): 971–983. Doi:10.1093/jexbot/51.347.971. Pmid 10948225
- Plant-growth, chapter, suny-biology2xmaster: lumenlearning.com, Retrieved 30 January, 2019
- Radoeva, tatyana; weijers, dolf (november 2014). "a roadmap to embryo identity in plants". Trends in plant science. 19 (11): 709–716. Doi:10.1016/j.tplants.2014.06.009. Pmid 25017700

Horticultural Practices and Techniques

3

- Nursery
- Pruning
- Plant Propagation
- Germination
- Tissue Culture
- Grafting
- Plant Breeding

There are a number of techniques which are used in horticulture. They include plant propagation, propagation, tissue culture, grafting and plant breeding. The diverse applications of these practices and techniques of horticulture have been thoroughly discussed in this chapter.

Nursery

Nursery is a place where seedlings, cuttings and grafts are raised with care before transplanting.

Advantage of Raising Seedlings in Nursery

1. It is very convenient to look after the tender seedlings.
2. It is easy to protect the seedlings from pests and diseases.
3. Economy of land usage (duration in the main field is reduced).
4. Valuable and very small seeds can be raised effectively without any wastage.
5. Uniform crop stand in the main field can be maintained by selecting healthy, uniform and vigorous seedlings in the nursery itself.

Preparation of Nursery Selection of Site

1. The nursery area should be nearer to the water source.
2. Generally, the location should be partially shaded i.e. under the trees. If not, artificial shade is to be provided.
3. It should be well protected from animals.
4. Proper drainage facilities should be provided.

Selection of Soil

A medium textured, loam (or) sand loam soil is preferred. Soil should be rich in organic matter. Soil depth should be preferably by 15-25 cm.

Types of Nursery Bed

1. Flat bed,
2. Raised nursery bed.

Preparation of Raised Nursery Bed

Selected soil should be worked well to break the clods. Weeds, stones and stubbles should be removed. Height of the raised bed should be 10-15 cm with a width of 1m and length may be according to the requirement and conveniences. Two parts of fine red earth, one part of sand and one part of FYM can be incorporated to each bed to improve aeration and fertility of the soil. Before preparing the bed, the soil should be drenched wit 4 % formaldehyde or 0.3 % copper oxy chloride to kill the pathogenic spores in the soil.

Advantage of Raised Nursery Bed

1. Water movement will be uniform and drainage of excess water is possible (In the case of flat bed water moves from one end to the other and there is possibility of washing away of seeds).
2. Germination percentage of seeds is normally high. Operations like weeding and plant protection measures are easy.

Media for Propagating Nursery Plants

Several materials and combination of different materials are available are media for germinating seeds and rooting cuttings. A good propagating medium should possess the following characters:

1. It must be firm and dense to hold the cuttings or seeds in place during rooting or germination.
2. It must possess sufficient moisture retaining capacity.

3. It must be sufficiently porous to permit excess water to drain away and to admit proper aeration.
4. It must be free from weed seeds, nematodes and pathogens.

Soil Mixture

This is the most commonly employed medium for pot plants. It usually consists of red earth, well decomposed cattle manure, leaf mold, river sand and also charcoal in some cases. Soil mixture commonly used for propagation is:

1. Red earth - 2 parts,
2. FYM - 1 part,
3. Sand - 1 part.

Sand

It is the most satisfactory medium for rooting of cuttings.

Peat

It consists of the remains of aquatic marsh, bog or swamp vegetation which has been preserved under water in a partially decomposed state. When such peat is derived from sphagnum, hypnum or other mosses, it is known as peat moss. it is used in mixture after breaking them and moistened.

Sphagnum Moss

Commercial sphagnum moss is the dehydrated young residue or living portion of acid-bog plants in the genus Sphagnum such as *S. papilliosum, S. capillacem* and *S. palustre.* it is generally collected from the tree trunks of the forest species in south Indian hills above 1500m above M.S.L. during rainy period. It is relatively sterile, light in weith and has a very high water-holding capacity. It is the commonly used medium in air layering.

Vermiculite

It is very light in weight and able to absorb large quantities of water. This can be used as a rooting medium for air layering and also in pots for raising certain plants.

Container for Propagation and Growing Young Plants

Earthen Pots

They are made of burnt porous clay in various sizes to provide requisite amount of soil and root space to different kinds and sizes of plants. They have straight sides and are made wider at the top than at the bottom to hold the greatest bulk of compost where the feeding roots are and also to facilitate easy removal of soil, intact with roots (ball of earth) at the time of planting or repotting.

In our county, tube pots of varying sizes are used as follows:

Tube pot sizes	Height (cm)	Diameter (cm)	Cost per pot (Rs.)
Tube pot	20	13	15.00
¼ size pot	18	22	15.00
½ size pot	20	27	30.00
¾ size pot	25	32	50.00
Full size pot	35	35	65.00
Tub size pots	35	50	90.00

Seed Pan and Seed Boxes

Seeds pans are shallow earthen pots about 10 cm high and 35 cm in diameter at the top. They have one large hole for drainage in the centre or 3 holes at equidistant from each other. Seed boxes are made of wood, 40 cm wide and 60 cm long and 10 cm deep, with 6-8 properly spaced holes drilled in the bottom.

Against each of the holes is placed a crock with its concave side down. Some large pieces of crock are put over it and also by the side of this crock, some coarse sand 2 or 3 handfuls are sprinkled on the crock pieces forming a thin layer to prevent fine soil from clogging the drainage. Over this, required soil mixture is added. Very delicate kinds of seeds like Cineraria, Begonia, etc. are best sown in these containers.

Polythene Bags

Small polythene bags with holes punched in the bottom for drainage and filled with a porous rooting medium are used for propagation of cuttings like Jasmines, Duranta, Crotons etc. in the mist chamber. Sometimes, young seedlings which are raised in the nursery are subsequently transplanted in these polythene bags and kept there till they attain required growth for transplanting them to the main field (Papaya, Curry leaf etc.).

Plastic Pots

Plastic pots, round and square are used to keep mostly indoor plants. They are reusable, light weight, non-porous and they require only little storage space.

Tools and Implements for Nursery Work

Rose can/water can	This is used for watering the nursery. Fine spray of water should be used for watering nursery of small sized seeds.
Digging fork	This has prongs of 20 cm long fitted to a wooden handle. This is used for uprooting plants, rooted cuttings, harvesting of tubers etc., without damaging the root system or tubers.
Shovel	This is a curved steel plate attached to a wooden handle and used for transferring soil, manure etc.
Garden rake	This is used for leveling lands and collecting weeds. The rake consists of a number of nail like projections from a crow bar provided with long handle.

Hand trowel	This is used as a small tool for making holes for planting seedlings and small plants. This is also useful for removing surface weeds in nursery beds.
Secateur	This is used for cutting small shoots to regulate shoot growth in fruit trees.
Budding or Grafting knife	This knife is used for budding and grafting. This has two blades in which one is with ivory edge used for lifting the bark in budding operation.

Potting

Purposes for which plants are potted are:

1. Preparing plants for sale such as rooted cuttings of grapes.
2. Growing plants for decoration like crotons.
3. Growing plants for experimental studies like pot -culture studies.
4. For using plants as rootstocks in certain grafting methods as in inarching of mango.

Pot Mixture or Potting Compost

It is essential for potting of plants. The pot mixture is prepared by suing various ingredients. The proportion of pot mixtures will vary with different kinds of plants.

1. An ideal pot mixture should have an open structure, which allows good drainage and holds sufficient moisture for plant growth and permits excess waster to drain away.
2. Should supply adequate nutrient to the plants during all stages of growth.
3. Should be free from all harmful organisms and toxic minerals.
4. Should be light in weight.

Potting Procedure

1. Wet the seedbed before lifting plants. Life with a ball of earth with as much of the root system intact, as possible. Do not pullout seedlings in the hot sun. Do not allow roots or the soil around the roots to dry.
2. Fill up pots by putting some crocks first, then a layer of sand (5-8 cm) and finally pot mixture (8-10 cm).
3. Place the plant with the ball of earth in the centre upon the layer of pot mixture (Place on one side of pots in the case of root stock plants used in inarching).
4. Put pot mixture around the ball of earth, press as you fill up and level off, leaving one inch head space at top. Do not press over the ball of earth. It will break and damage the roots.
5. Set the stem of plant at the same height as it was in the seed bed.
6. Immerse pot with plant in a tub of water gently and keep inside water till air bubbles cease to come out. Remove and place the pot under shade of trees.

Repotting

Repotting is done for changing the soil medium for pot bound plants.

Pot Bound Condition

When the potted plants are grown for more than one season or one year in pot, the root very soon become a tangled mass and exhaust all the nutrient in the limited soil, besides being circumscribed in the limited place. This stage is known as pot bound condition.

Repotting Procedure

1. It is better to west the potted plant 24 hours earlier to facilitate repotting (removal of plant from pot).
2. The technique to remove the plant with a ball')f each intact is to keep the right hand palm over the soil, allowing the stem of the plant in between the first fingers and turn the pot upside down holding the pot at the bottom with the left hand and gently knocking the rim of the pot on the edge of table or any other hard surface or even on the bottom edge of another inverted pot. The ball of earth comes out of the pot. If for any reason, it fails to come out, break the pot knocking the sides with a stone or fork and free the soil from it.
3. Examine the roots, cut neatly with a secateur, the decayed, dead and dried or twisted roots. Reduce the size of the ball of earth around the roots.
4. Place the plant in the new pot at the same height at which it was in the old pot. Fill up pot with fresh pot mixture and immerse in water.
5. The initial reaction after potting and repotting is wilting. The transpiration loss has to be checked to help plants revive. Hence keep freshly potted plants under shade and "pot water daily".
6. After about ten days under shade, the plants should be gradually exposed to sun by keeping them for some hours under sun and then putting them under shade. The period of exposure can be increased every week until finally the plants can be kept in the open. This process is called "hardening".

Pruning

Pruning is a horticultural and silvicultural practice involving the selective removal of certain parts of a plant, such as branches, buds, or roots. Reasons to prune plants include deadwood removal, shaping (by controlling or redirecting growth), improving or sustaining health, reducing risk from falling branches, preparing nursery specimens for transplanting, and both harvesting and increasing the yield or quality of flowers and fruits.

The practice entails targeted removal of diseased, damaged, dead, non-productive, structurally unsound, or otherwise unwanted tissue from crop and landscape plants. In general, the smaller the branch that is cut, the easier it is for a woody plant to compartmentalize the wound and thus limit the potential for pathogen intrusion and decay. It is therefore preferable to make any necessary formative structural pruning cuts to young plants, rather than removing large, poorly placed branches from mature plants.

Pruning: dense growth after shearing.

Specialized pruning practices may be applied to certain plants, such as roses, fruit trees, and grapevines. It is important when pruning that the tree's limbs are kept intact, as this is what helps the tree stay upright. Different pruning techniques may be deployed on herbaceous plants than those used on perennial woody plants. Hedges, by design, are usually (but not exclusively) maintained by hedge trimming, rather than by pruning.

In nature, meteorological conditions such as wind, ice and snow, and salinity can cause plants to self-prune. This natural shedding is called abscission.

Pruning Landscape and Amenity Trees

Types of Branch Union

Pruning when there's a branch collar: The swollen area where the branch joins the trunk is known as the colla.

For arboricultural purposes the unions of tree branches (i.e. where they join together) are placed in one of three types: collared, collarless or codominant. Regardless of the overall type of pruning being carried out, each type of union is cut in a particular way so that the branch has less chance of regrowth from the cut area and best chance of sealing over and compartmentalising decay. This is often referred to by arborists as "target cutting".

Deadwooding

Branches die off for a number of reasons including light deficiency, pest and disease damage, and root structure damage. A dead branch will at some point decay back to the parent stem and fall off. This is normally a slow process but can be quickened by high winds or extreme temperature. The main reason deadwooding is performed is safety. Situations that usually demand removal of deadwood is trees that overhang public roads, houses, public areas and gardens. Trees located in wooded areas are usually assessed as lower risk but assessments consider the number of visitors. Usually, trees adjacent to footpaths and access roads are considered for deadwood removal.

Another reason for deadwooding is amenity value, i.e. a tree with a large amount of deadwood throughout the crown looks more aesthetically pleasing with the deadwood removed. The physical practice of deadwooding can be carried out most of the year though not when the tree is coming into leaf. The deadwooding process speeds up the tree's natural abscission process. It also reduces unwanted weight and wind resistance and can help overall balance.

Crown and Canopy Thinning

Crown and canopy thinning increases light and reduces wind resistance by selective removal of branches throughout the canopy of the tree.

Crown Canopy Lifting

Crown lifting involves the removal of the lower branches to a given height. The height is achieved by the removal of whole branches or removing the parts of branches which extend below the desired height. The branches are normally not lifted to more than one third of the tree's total height.

Pruning when there's a collarless union: The area where there is no swollen area where the branch joins the trunk is called a collarless union.

Crown lifting is done for access; these being pedestrian, vehicle or space for buildings and street furniture. Lifting the crown will allow traffic and pedestrians to pass underneath safely. This pruning technique is usually used in the urban environment as it is for public safety and aesthetics rather than tree form and timber value.

Crown lifting introduces light to the lower part of the trunk; this, in some species can encourage epicormic growth from dormant buds. To reduce this sometimes smaller branches are left on the lower part of the trunk. Excessive removal of the lower branches can displace the canopy weight, this will make the tree top heavy, therefore adding stress to the tree. When a branch is removed from the trunk, it creates a large wound. This wound is susceptible to disease and decay, and could lead to reduced trunk stability. Therefore, much time and consideration must be taken when choosing the height the crown is to be lifted to.

This would be an inappropriate operation if the tree species' form was of a shrubby nature. This would therefore remove most of the foliage and would also largely unbalance the tree. This procedure should not be carried out if the tree is in decline, poor health or dead, dying or dangerous (DDD) as the operation will remove some of the photosynthetic area the tree uses. This will increase the decline rate of the tree and could lead to death.

If the tree is of great importance to an area or town, (i.e. veteran or ancient) then an alternative solution to crown lifting would be to move the target or object so it is not in range. For example, diverting a footpath around a tree's drip line so the crown lift is not needed. Another solution would be to prop up or cable-brace the low hanging branch. This is a non-invasive solution which in some situations can work out more economically and environmentally friendly.

Pruning when it's a codominant stem.

Directional or Formative Pruning

Removal of appropriate branches to make the tree structurally sound while shaping it.

Vista Pruning

Selectively pruning a window of view in a tree.

Crown Reduction

Reducing the height and or spread of a tree by selectively cutting back to smaller branches and in fruit trees for increasing of light interception and enhancing fruit quality.

Pollarding

A regular form of pruning where certain deciduous species are pruned back to pollard heads every year in the dormant period. This practice is usually commenced on juvenile trees so they can adapt to the harshness of the practice.

Types

Arborists, orchardists, and gardeners use various garden tools and tree cutting tools designed for the purpose, such as hand pruners, loppers, or chainsaws. Additionally in forestry bush saws are commonly used and these are often attached to poles that reach up to 5-6m, this is a more efficient way of pruning than with ladders. These bush saws on polls have also been motorized as chainsaws which is even more efficient. Older technology used Billhooks, Kaiser blades and pruning knives. Although still used in some coppicing they are not used so much in commercial forestry due to the difficulty of cutting flush with the stem. Flush cuts lead to good wood, non-flush or bark damaging cuts (which are more likely with a swung blade than a sawed one) put the tree at risk of entry cords from forest pathogens.

Regardless of the various names used for types of pruning, there are only two basic cuts: One cuts back to an intermediate point, called heading back cut, and the other cuts back to some point of origin, called thinning out cut.

Removing a portion of a growing stem down to a set of desirable buds or side-branching stems. This is commonly performed in well trained plants for a variety of reasons, for example to stimulate growth of flowers, fruit or branches, as a preventative measure to wind and snow damage on long stems and branches, and finally to encourage growth of the stems in a desirable direction. Also commonly known as heading-back.

Thinning: A more drastic form of pruning, a thinning out cut is the removal of an entire shoot, limb, or branch at its point of origin. This is usually employed to revitalize a plant by removing over-mature, weak, problematic, and excessive growths. When performed correctly, thinning encourages the formation of new growth that will more readily bear fruit and flowers. This is a common technique in pruning roses and for amplifying and "opening-up" the branching of neglected trees, or for renewing shrubs with multiple branches.

Topping: Topping is a very severe form of pruning which involves removing all branches and growths down to a few large branches or to the trunk of the tree. When performed correctly it is used on very young trees, and can be used to begin training younger trees for pollarding or for trellising to form an espalier.

Raising removes the lower branches from a tree in order to provide clearance for buildings, vehicles, pedestrians, and vistas.

Reduction reduces the size of a tree, often for clearance for utility lines. Reducing the height or spread of a tree is best accomplished by pruning back the leaders and branch terminals to lateral branches that are large enough to assume the terminal roles (at least one-third the diameter of the cut stem). Compared to topping, reduction helps maintain the form and structural integrity of the tree.

In orchards, fruit trees are often lopped to encourage regrowth and to maintain a smaller tree for ease of picking fruit. The pruning regime in orchards is more planned and the productivity of each tree is an important factor.

Deadheading is the act of removing spent flowers or flowerheads for aesthetics, to prolong bloom for up to several weeks or promote rebloom, or to prevent seeding.

Time Period

In general, pruning dead wood and small branches can be done at any time of year. Depending on the species, many temperate plants can be pruned either during dormancy in winter, or, for species where winter frost can harm a recently pruned plant, after flowering is completed. In the temperate areas of the northern hemisphere autumn pruning should be avoided, as the spores of disease and decay fungi are abundant at this time of year.

Some woody plants tend to bleed profusely from cuts, such as mesquite and maple. Some callus over slowly, such as magnolia. In this case, they are better pruned in during active growth when they can more readily heal. Woody plants that flower early in the season, on spurs that form on wood that has matured the year before, such as apples, should be pruned right after flowering as later pruning will sacrifice flowers the following season. Forsythia, azaleas and lilacs all fall into this category.

Pollarding

Pollarding, a pruning system involving the removal of the upper branches of a tree, promotes a dense head of foliage and branches. In ancient Rome, Propertius mentioned pollarding during the 1st century BCE. The practice occurred commonly in Europe since medieval times, and takes place today in urban areas worldwide, primarily to maintain trees at a determined height.

Traditionally, people pollarded trees for one of two reasons: for fodder to feed livestock or for wood. Fodder pollards produced "pollard hay" for livestock feed; they were pruned at intervals of two to six years so their leafy material would be most abundant. Wood pollards were pruned at longer intervals of eight to fifteen years, a pruning cycle tending to produce upright poles favored for fence rails and posts and boat construction. Supple young willow or hazel branches may be harvested as material for weaving baskets, fences, and garden constructions such as bowers. Nowadays, the practice is sometimes used for ornamental trees, such as crepe myrtles in southern states of the USA, although the resulting tree has a stunted form rather than a natural-looking crown.

Pollarding tends to make trees live longer by maintaining them in a partially juvenile state and by reducing the weight and windage of the top part of the tree. Older pollards often become hollow, so can be difficult to age accurately. Pollards tend to grow slowly, with denser growth-rings in the years immediately after cutting.

Practice

Pollarding began with walled cities in Europe which did not have room for large trees. The smaller limbs that resulted could be used for heat and cooking.

As in coppicing, pollarding is to encourage the tree to produce new growth on a regular basis to maintain a supply of new wood for various purposes, particularly for fuel. In some areas, dried leafy branches are stored as winter fodder for stock. Depending on the use of the cut material, the length of time between cutting will vary from one year for tree hay or withies, to five years or more for larger timber. Sometimes, only some of the regrown stems may be cut in a season – this is thought to reduce the chances of death of the tree when recutting long-neglected pollards.

Newly pollarded willow trees between Sluis and Aardenburg in Zeeland.

The same trees two years later, showing the gradual expansion of the trees' crowns.

Pollarding was preferred over coppicing in wood-pastures and other grazed areas, because animals would browse the regrowth from coppice stools. Historically, the right to pollard or "lop" was often granted to local people for fuel on common land or in royal forests; this was part of the right of Estover.

An incidental effect of pollarding in woodland is the encouragement of underbrush growth due to increased light reaching the woodland floor. This can increase species diversity. However, in woodland where pollarding was once common but has now ceased, the opposite effect occurs, as the side and top shoots develop into trunk-sized branches. An example of this can be seen in Epping Forest in London/Essex, UK, the majority of which was pollarded until the late 19th century. Here, the light that reaches the woodland floor is extremely limited owing to the thick growth of the pollarded trees.

Pollards cut at about a metre above the ground are called stubs (or stubbs). These were often used as markers in coppice or other woodland. Stubs cannot be used where the trees are browsed by animals, as the regrowing shoots are below the browse line.

Although people who migrated to the United States from Europe continued the practice, experts have come to believe that pollarding older trees harms the tree. The smaller limbs grow from wood that is not as strong, and the weaker trees will not live as long and can be more easily damaged by storms.

Species

As with coppicing, only species with vigorous epicormic growth may be made into pollards. In these species (which include many broadleaved trees but few conifers), removal of the main apical stems releases the growth of many dormant buds under the bark on the lower part of the tree. Trees without this growth will die without their leaves and branches. Some smaller tree species do not readily form pollards, because cutting the main stem stimulates growth from the base, effectively forming a coppice stool instead. Examples of trees that do well as pollards include broadleaves such as beeches (Fagus), oaks (Quercus), maples (Acer), black locust or false acacia (Robinia pseudoacacia), hornbeams (Carpinus), lindens and limes (Tilia), planes (Platanus), horse chestnuts (Aesculus), mulberries (Morus), Eastern redbud (Cercis canadensis), tree of heaven (Ailanthus altissima), willows (Salix), and a few conifers, such as yews (Taxus).

Ancient pollarded beech tree in Epping Forest.

The technique is used in Africa for moringa trees to bring the nutritious leaves into easier reach for harvesting. Pollarding is also used in urban forestry in certain areas for reasons such as tree size management, safety, and health concerns. It removes rotting or diseased branches to support the overall health of the tree and removes living and dead branches that could harm property and people, as well as increasing the amount of foliage in spring for aesthetic, shade and air quality reasons. Some trees may be rejuvenated by pollarding – for example, Bradford pear (Pyrus calleryana "Bradford"), a beautiful flowering species when young that becomes brittle and top-heavy when older.

Oaks, when very old, can form new trunks from the growth of pollard branches, i.e. surviving branches which have split away from the main branch naturally.

Plant Propagation

Propagation, in horticulture is the reproduction of plants by any number of natural or artificial means.

Sexual Propagation

With crops that produce seed freely and come true closely enough for the purposes in view, growing from seed usually is the cheapest and most satisfactory method of plant propagation. Many types of seeds may be sown in open ground and, barring extreme wetness or extreme aridity, germinate well enough for practical purposes. Other kinds, however, are so exacting in their requirements that these are best met in a propagating house where humidity and temperature can be more rigidly controlled. Because of their high oxygen requirement, the medium in which the seeds are sown generally should contain more sand (or other filler or mulch material) than ordinary garden soil does. Greater porosity makes these media more subject to rapid drying, however, and moisture must be carefully monitored. Because many soils harbour fungi destructive to sprouting seed and young seedlings, soil that is used for germinating seed commonly is sterilized by heat or chemicals. Many diseases of plants are caused by fungi and bacteria carried in or on the seed itself, and treatment of the seed with disinfectants is beneficial.

Asexual Propagation

Some species of plants, in their cultivated forms, do not produce seed—e.g., banana, pineapple, and sugarcane. In a great number of cultivated species, seedlings vary so much that the desired traits are found in only a small proportion. For these and other reasons, horticulturists resort to asexual propagation—i.e., the division or separation and indefinite subdivision of the original plant having the desired traits.

Many people have held the opinion that asexual propagation is unnatural and that plants thus derived lack the hardiness or the sturdiness of plants grown from seed. Asexual propagation, however, is not unnatural; some of its forms—e.g., layering and grafting—are rather common in nature. The only justification for the generalization that asexually propagated plants lack hardiness or sturdiness is the fact that virus diseases can be transmitted by asexual propagation and that most such diseases cannot be transmitted through seed.

Methods of asexual propagation include bulb division, layering, cutting, and grafting. Bulbs and other underground rootlike structures, such as tubers and corms, may be divided as they mature. The sections are then placed in a moist medium to root. In layering, the stem of a large plant is notched and wrapped in moist sphagnum moss or bent to the ground and covered with moist soil; when roots appear, growing out of the moss, the stem is cut below the roots and potted. Auxins (growth hormones) are often added to the wounds or soil to stimulate rooting. Stem cuttings are rooted in water or a moist potting medium such as sand, peat moss, or vermiculite. When cutting or layering are not feasible, a bud or twig of one plant is grafted onto the fully developed root system of another.

Germination

Germination refers to the process by which an organism grows from a seed or a spore. The most common forms of germination include a seed sprouting to form a seedling and the formation of a sporeling from a spore. Thus, germination occurs primarily in plant and fungal species.

Germination Process

The process of germination is as follows:

1. Environmental conditions are favorable: For germination to occur, the environmental conditions must be favorable in order to support the growing plant. The soil depth, amount of water, and temperature are all critical conditions that must be met in order for the process of germination to be initiated. Typically, the soil conditions must be moist and warm.
2. Water imbibition: When environmental conditions are optimal, germination is initiated by a process termed water imbibition. The seed absorbs water through a structure called a micropyle, which induces swelling of the seed until it splits open.
3. Root and Shoot formation: Once the seed has ruptured, the radicle (primary root) and plumule (shoot) can emerge from the seed. This process is initiated by specific enzymes that become activated when the seed is exposed to water. The roots grow downwards, and the shoot grows upwards towards the soil surface.
4. A seedling is formed: Once the shoot emerges from the soil surface, the cotyledons become fully unfolded and expand, eventually forming the first leaves. Once this occurs, the plant is ready to initiate photosynthesis and is considered a seedling.

Germination Temperature

Among other conditions, the temperature is critical for germination to occur. Although the temperature will affect the plant's growth rate and metabolism, most plants will germinate over a wide temperature range (e.g., 16 to 24 degrees Celsius). Depending on the specific climate, some plants will only germinate when conditions are cool, while others require warm temperatures. Moreover, there are several plant species that require temperatures to fluctuate between cold and warm in order to break dormancy and facilitate germination. For example, some seeds require exposure to cold winter temperatures (e.g., 4 to -5 degrees Celsius) prior to germination, while other require extreme heat (e.g., forest fires) to crack the seed in order to initiate germination. It is thought that it is a mechanism by which optimal growth conditions can be anticipated. For example, some seeds absorb water during the fall, which causes the seed to erupt during the cold winter temperatures. Once the soil warms again, the sprout and roots will emerge, forming a seedling just as the amount of water and sunlight increase during the spring and summer months.

Tissue Culture

Plant tissue culture is a collection of techniques used to maintain or grow plant cells, tissues or organs under sterile conditions on a nutrient culture medium of known composition. Plant tissue culture is widely used to produce clones of a plant in a method known as micropropagation. Different techniques in plant tissue culture may offer certain advantages over traditional methods of propagation, including:

- The production of exact copies of plants that produce particularly good flowers, fruits, or have other desirable traits.
- To quickly produce mature plants.
- The production of multiples of plants in the absence of seeds or necessary pollinators to produce seeds.
- The regeneration of whole plants from plant cells that have been genetically modified.
- The production of plants in sterile containers that allows them to be moved with greatly reduced chances of transmitting diseases, pests, and pathogens.
- The production of plants from seeds that otherwise have very low chances of germinating and growing, i.e.: orchids and Nepenthes.
- To clear particular plants of viral and other infections and to quickly multiply these plants as 'cleaned stock' for horticulture and agriculture.

Plant tissue culture relies on the fact that many plant cells have the ability to regenerate a whole plant (totipotency). Single cells, plant cells without cell walls (protoplasts), pieces of leaves, stems or roots can often be used to generate a new plant on culture media given the required nutrients and plant hormones.

Regeneration Pathways

Plant tissue cultures being grown at a USDA seed bank, the National Center for Genetic Resources Preservation.

The specific differences in the regeneration potential of different organs and explants have various explanations. The significant factors include differences in the stage of the cells in the cell cycle, the availability of or ability to transport endogenous growth regulators, and the metabolic capabilities of the cells. The most commonly used tissue explants are the meristematic ends of the plants like the stem tip, axillary bud tip and root tip. These tissues have high rates of cell division and either concentrate or produce required growth regulating substances including auxins and cytokinins.

Shoot regeneration efficiency in tissue culture is usually a quantitative trait that often varies between plant species and within a plant species among subspecies, varieties, cultivars, or ecotypes. Therefore, tissue culture regeneration can become complicated especially when many regeneration procedures have to be developed for different genotypes within the same species.

The three common pathways of plant tissue culture regeneration are propagation from preexisting meristems (shoot culture or nodal culture), organogenesis and non-zygotic embryogenesis.

The propagation of shoots or nodal segments is usually performed in four stages for mass production of plantlets through in vitro vegetative multiplication but organogenesis is a common method of micropropagation that involves tissue regeneration of adventitious organs or axillary buds directly or indirectly from the explants. Non-zygotic embryogenesis is a noteworthy developmental pathway that is highly comparable to that of zygotic embryos and it is an important pathway for producing somaclonal variants, developing artificial seeds, and synthesizing metabolites. Due to the single cell origin of non-zygotic embryos, they are preferred in several regeneration systems for micropropagation, ploidy manipulation, gene transfer, and synthetic seed production. Nonetheless, tissue regeneration via organogenesis has also proved to be advantageous for studying regulatory mechanisms of plant development.

Choice of Explant

The tissue obtained from a plant to be cultured is called an explant.

Explants can be taken from many different parts of a plant, including portions of shoots, leaves, stems, flowers, roots, single undifferentiated cells and from many types of mature cells provided are they still contain living cytoplasm and nuclei and are able de-differentiate and resume cell division. This has given rise to the concept of totipotentency of plant cells. However this is not true for all cells or for all plants. In many species explants of various organs vary in their rates of growth and regeneration, while some do not grow at all. The choice of explant material also determines if the plantlets developed via tissue culture are haploid or diploid. Also the risk of microbial contamination is increased with inappropriate explants.

The first method involving the meristems and induction of multiple shoots is the preferred method for the micropropagation industry since the risks of somaclonal variation (genetic variation induced in tissue culture) are minimal when compared to the other two methods. Somatic embryogenesis is a method that has the potential to be several times higher in multiplication rates and is amenable to handling in liquid culture systems like bioreactors.

Some explants, like the root tip, are hard to isolate and are contaminated with soil microflora that become problematic during the tissue culture process. Certain soil microflora can form tight associations with the root systems, or even grow within the root. Soil particles bound to roots are difficult to remove without injury to the roots that then allows microbial attack. These associated microflora will generally overgrow the tissue culture medium before there is significant growth of plant tissue.

Some cultured tissues are slow in their growth. For them there would be two options: (i) Optimizing the culture medium; (ii) Culturing highly responsive tissues or varieties. Necrosis can spoil cultured tissues. Generally, plant varieties differ in susceptibility to tissue culture necrosis. Thus, by culturing highly responsive varieties (or tissues) it can be managed.

Aerial (above soil) explants are also rich in undesirable microflora. However, they are more easily removed from the explant by gentle rinsing, and the remainder usually can be killed by surface sterilization. Most of the surface microflora do not form tight associations with the plant tissue. Such associations can usually be found by visual inspection as a mosaic, de-colorization or localized necrosis on the surface of the explant.

An alternative for obtaining uncontaminated explants is to take explants from seedlings which are aseptically grown from surface-sterilized seeds. The hard surface of the seed is less permeable to penetration of harsh surface sterilizing agents, such as hypochlorite, so the acceptable conditions of sterilization used for seeds can be much more stringent than for vegetative tissues.

Tissue cultured plants are clones. If the original mother plant used to produce the first explants is susceptible to a pathogen or environmental condition, the entire crop would be susceptible to the same problem. Conversely, any positive traits would remain within the line also.

Applications

Plant tissue culture is used widely in the plant sciences, forestry, and in horticulture. Applications include:

- The commercial production of plants used as potting, landscape, and florist subjects, which uses meristem and shoot culture to produce large numbers of identical individuals.
- To conserve rare or endangered plant species.
- A plant breeder may use tissue culture to screen cells rather than plants for advantageous characters, e.g. herbicide resistance/tolerance.

- Large-scale growth of plant cells in liquid culture in bioreactors for production of valuable compounds, like plant-derived secondary metabolites and recombinant proteins used as biopharmaceuticals.
- To cross distantly related species by protoplast fusion and regeneration of the novel hybrid.
- To rapidly study the molecular basis for physiological, biochemical, and reproductive mechanisms in plants, for example in vitro selection for stress tolerant plants.
- To cross-pollinate distantly related species and then tissue culture the resulting embryo which would otherwise normally die (Embryo Rescue).
- For chromosome doubling and induction of polyploidy, for example doubled haploids, tetraploids, and other forms of polyploids. This is usually achieved by application of antimitotic agents such as colchicine or oryzalin.
- As a tissue for transformation, followed by either short-term testing of genetic constructs or regeneration of transgenic plants.
- Certain techniques such as meristem tip culture can be used to produce clean plant material from virused stock, such as sugarcane, potatoes and many species of soft fruit.
- Production of identical sterile hybrid species can be obtained.
- Large scale production of artificial seeds through somatic embryogenesis.

Grafting

Grafting or graftage is a horticultural technique whereby tissues of plants are joined so as to continue their growth together. The upper part of the combined plant is called the scion while the lower part is called the rootstock. The success of this joining requires that the vascular tissue grow together and such joining is called inosculation. The technique is most commonly used in asexual propagation of commercially grown plants for the horticultural and agricultural trades.

In most cases, one plant is selected for its roots and this is called the stock or rootstock. The other plant is selected for its stems, leaves, flowers, or fruits and is called the scion or cion. The scion contains the desired genes to be duplicated in future production by the stock/scion plant.

In stem grafting, a common grafting method, a shoot of a selected, desired plant cultivar is grafted onto the stock of another type. In another common form called bud grafting, a dormant side bud is grafted onto the stem of another stock plant, and when it has inosculated successfully, it is encouraged to grow by pruning off the stem of the stock plant just above the newly grafted bud.

For successful grafting to take place, the vascular cambium tissues of the stock and scion plants must be placed in contact with each other. Both tissues must be kept alive until the graft has "taken", usually a period of a few weeks. Successful grafting only requires that a vascular connection take place between the grafted tissues. Research conducted in Arabidopsis thaliana hypocotyls have shown that the connection of phloem takes place after 3 days of initial grafting, whereas the connection of xylem can take up to 7 days. Joints formed by grafting are not as strong as naturally formed joints, so a physical weak point often still occurs at the graft because only the newly formed tissues inosculate with each other. The existing structural tissue (or wood) of the stock plant does not fuse.

Advantages

Graft particular to plum cherry: The scion is the largest in the plant, due to the imperfect union of the two. It can be seen on the enlarged trunk: this accumulation of starch is an indication of imperfection.

- Precocity: The ability to induce fruitfulness without the need for completing the juvenile phase. Juvenility is the natural state through which a seedling plant must pass before it can become reproductive. In most fruiting trees, juvenility may last between 5 and 9 years, but in some tropical fruits e.g. Mangosteen, juvenility may be prolonged for up to 15 years. Grafting of mature scions onto rootstocks can result in fruiting in as little as two years.
- Dwarfing: To induce dwarfing or cold tolerance or other characteristics to the scion. Most apple trees in modern orchards are grafted on to dwarf or semi-dwarf trees planted at high density. They provide more fruit per unit of land, higher quality fruit, and reduce the danger of accidents by harvest crews working on ladders. Care must be taken when planting dwarf or semi-dwarf trees. If such a tree is planted with the graft below the soil, then the scion portion can also grow roots and the tree will still grow to its standard size.
- Ease of propagation: Because the scion is difficult to propagate vegetatively by other means, such as by cuttings. In this case, cuttings of an easily rooted plant are used to provide a rootstock. In some cases, the scion may be easily propagated, but grafting may still be used because it is commercially the most cost-effective way of raising a particular type of plant.
- Hybrid breeding: To speed maturity of hybrids in fruit tree breeding programs. Hybrid seedlings may take ten or more years to flower and fruit on their own roots. Grafting can reduce the time to flowering and shorten the breeding program.
- Hardiness: Because the scion has weak roots or the roots of the stock plants are tolerant of difficult conditions. e.g. many Western Australian plants are sensitive to dieback on heavy soils, common in urban gardens, and are grafted onto hardier eastern Australian relatives. Grevilleas and eucalypts are examples.

- Sturdiness: To provide a strong, tall trunk for certain ornamental shrubs and trees. In these cases, a graft is made at a desired height on a stock plant with a strong stem. This is used to raise 'standard' roses, which are rose bushes on a high stem, and it is also used for some ornamental trees, such as certain weeping cherries.
- Disease/pest resistance: In areas where soil-borne pests or pathogens would prevent the successful planting of the desired cultivar, the use of pest/disease tolerant rootstocks allow the production from the cultivar that would be otherwise unsuccessful. A major example is the use of rootstocks in combating Phylloxera.
- Pollen source: To provide pollenizers. For example, in tightly planted or badly planned apple orchards of a single variety, limbs of crab apple may be grafted at regularly spaced intervals onto trees down rows, say every fourth tree. This takes care of pollen needs at blossom time, yet does not confuse pickers who might otherwise mix varieties while harvesting, as the mature crab apples are so distinct from other apple varieties.
- Repair: To repair damage to the trunk of a tree that would prohibit nutrient flow, such as stripping of the bark by rodents that completely girdles the trunk. In this case a bridge graft may be used to connect tissues receiving flow from the roots to tissues above the damage that have been severed from the flow. Where a water sprout, basal shoot or sapling of the same species is growing nearby, any of these can be grafted to the area above the damage by a method called inarch grafting. These alternatives to scions must be of the correct length to span the gap of the wound.
- Changing cultivars: To change the cultivar in a fruit orchard to a more profitable cultivar, called *top working*. It may be faster to graft a new cultivar onto existing limbs of established trees than to replant an entire orchard.
- Maintain consistency: Apples are notorious for their genetic variability, even differing in multiple characteristics, such as, size, color, and flavor, of fruits located on the same tree. In the commercial farming industry, consistency is maintained by grafting a scion with desired fruit traits onto a hardy stock.

An example of approach grafting by Axel Erlandson.

- Curiosities:
    - A practice sometimes carried out by gardeners is to graft related potatoes and tomatoes so that both are produced on the same plant, one above ground and one underground.
    - Cacti of widely different forms are sometimes grafted on to each other.
    - Multiple cultivars of fruits such as apples are sometimes grafted on a single tree. This so-called "family tree" provides more fruit variety for small spaces such as a suburban backyard, and also takes care of the need for pollenizers. The drawback is that the gardener must be sufficiently trained to prune them correctly, or one strong variety will usually "take over." Multiple cultivars of different "stone fruits" (*Prunus* species) can be grafted on a single tree. This is called a "fruit salad tree".
    - Ornamental and functional, tree shaping uses grafting techniques to join separate trees or parts of the same tree to itself. Furniture, hearts, entry archways are examples. Axel Erlandson was a prolific tree shaper who grew over 75 mature specimens.

Factors for Successful Graft

Compatibility of scion and stock: Because grafting involves the joining of vascular tissues between the scion and rootstock, plants lacking vascular cambium, such as monocots, cannot normally be grafted. As a general rule, the closer two plants are genetically, the more likely the graft union will form. Genetically identical clones and intra-species plants have a high success rate for grafting. Grafting between species of the same genus is sometimes successful. Grafting has a low success rate when performed with plants in the same family but in different genera. And grafting between different families is rare.

- Cambium alignment and pressure: The vascular cambium of the scion and stock should be tightly pressed together and oriented in the direction of normal growth. Proper alignment and pressure encourages the tissues to join quickly, allowing nutrients and water to transfer from the stockroot to the scion.
- Completed during appropriate stage of plant: The grafting is completed at a time when the scion and stock are capable of producing callus and other wound-response tissues. Generally, grafting is performed when the scion is dormant, as premature budding can drain the grafting site of moisture before the grafting union is properly established. Temperature greatly affects the physiological stage of plants. If the temperature is too warm, premature budding may result. Elsewise, high temperatures can slow or halt callus formation.
- Proper care of graft site: After grafting, it is important to nurse the grafted plant back to health for a period of time. Various grafting tapes and waxes are used to protect the scion and stock from excessive water loss. Furthermore, depending on the type of graft, twine or string is used to add structural support to the grafting site. Sometimes it is necessary to prune the site, as the rootstock may produce shoots that inhibit the growth of the scion.

Tools

Depiction of general purpose grafting knives:

- Cutting tools: It is good procedure to keep the cutting tool sharp to minimize tissue damage and clean from dirt and other substances to avoid the spread of disease. A good knife for general grafting should have a blade and handle length of about 3 inches and 4 inches respectively. Specialized knives for grafting include bud-grafting knives, surgical knives, and pruning knives. Cleavers, chisels, and saws are utilized when the stock is too large to be cut otherwise.
- Disinfecting tools: Treating the cutting tools with disinfectants ensures the grafting site is clear of pathogens. A common sterilizing agent is absolute alcohol.
- Graft seals: Keeps the grafting site hydrated. Good seals should be tight enough to retain moisture while, at the same time, loose enough to accommodate plant growth. Includes specialized types of clay, wax, petroleum jelly, and adhesive tape.
- Tying and support materials: Adds support and pressure to the grafting site to hold the stock and scion together before the tissues join, which is especially important in herbaceous grafting. The employed material is often dampened before use to help protect the site from desiccation. Support equipment includes strips made from various substances, twine, nails, and splints.
- Grafting machines: Because grafting can take a lot of time and skill, grafting machines have been created. Automation is particularly popular for seedling grafting in countries such as Japan and Korea where farming land is both limited and used intensively. Certain machines can graft 800 seedlings per hour.

Techniques

Approach

T budding.

Approach grafting or inarching is used to join together plants that are otherwise difficult to join. The plants are grown close together, and then joined so that each plant has roots below and growth above the point of union. Both scion and stock retain their respective parents that may or may not be removed after joining. Also used in pleaching. The graft can be successfully accomplished any time of year.

Bud

Bud grafting (also called chip budding) uses a bud instead of a twig. Grafting roses is the most common example of bud grafting. In this method a bud is removed from the parent plant, and the base of the bud is inserted beneath the bark of the stem of the stock plant from which the rest of the shoot has been cut. Any extra bud that starts growing from the stem of the stock plant is removed. Examples: roses and fruit trees like peaches.

Budwood is a stick with several buds on it that can be cut out and used for bud grafting. It is a common method of propagation for citrus trees.

Cleft

Successful cleft graft after 2 years growth.

Same graft after 4 years growth.

In cleft grafting a small cut is made in the stalk and then the pointed end of the scion is inserted in the stalk. This is best done in the early spring and is useful for joining a thin scion about 1 cm (3⁄8 in) diameter to a thicker branch or stock. It is best if the former has 3–5 buds and the latter is 2–7 cm (3⁄4–2 3⁄4 in) in diameter. The branch or stock should be split carefully down the middle to form a cleft about 3 cm (1 1⁄8 in) deep. If it is a branch that is not vertical then the cleft should be cut horizontally. The end of the scion should be cut cleanly to a long shallow wedge, preferably with a single cut for each wedge surface, and not whittled. A third cut may be made across the end of the wedge to make it straight across.

Slide the wedge into the cleft so that it is at the edge of the stock and the centre of the wedge faces are against the cambium layer between the bark and the wood. It is preferable if a second scion is inserted in a similar way into the other side of the cleft. This helps to seal off the cleft. Tape around the top of the stock to hold the scion in place and cover with grafting wax or sealing compound. This stops the cambium layers from drying out and also prevents the ingress of water into the cleft.

Whip

Graft needing attention the following season.

In whip grafting the scion and the stock are cut slanting and then joined. The grafted point is then bound with tape and covered with a soft sealant to prevent dehydration and infection by germs. The common variation is a whip and tongue graft, which is considered the most difficult to master but has the highest rate of success as it offers the most cambium contact between the scion and the stock. It is the most common graft used in preparing commercial fruit trees. It is generally used with stock less than 1.25 cm (1⁄2 in) diameter, with the ideal diameter closer to 1 cm (3⁄8 in) and the scion should be of roughly the same diameter as the stock.

The stock is cut through on one side only at a shallow angle with a sharp knife. (If the stock is a branch and not the main trunk of the rootstock then the cut surface should face outward from the centre of the tree.) The scion is similarly sliced through at an equal angle starting just below a bud, so that the bud is at the top of the cut and on the other side than the cut face.

Successful whip graft. In the whip and tongue variation, a notch is cut downwards into the sliced face of the stock and a similar cut upwards into the face of the scion cut. These act as the tongues and it requires some skill to make the cuts so that the scion and the stock marry up neatly. The elongated "Z" shape adds strength, removing the need for a companion rod in the first season.

The joint is then taped around and treated with tree-sealing compound or grafting wax. A whip graft without a tongue is less stable and may need added support.

Stub

Successful stub graft, healed.

Stub grafting is a technique that requires less stock than cleft grafting, and retains the shape of a tree. Also scions are generally of 6–8 buds in this process.

An incision is made into the branch 1 cm (3⁄8 in) long, then the scion is wedged and forced into the branch. The scion should be at an angle of at most 35° to the parent tree so that the crotch remains strong. The graft is covered with grafting compound.

After the graft has taken, the branch is removed and treated a few centimeters above the graft, to be fully removed when the graft is strong.

Four-flap

The four-flap graft (also called banana graft) is commonly used for pecans, and first became popular with this species in Oklahoma in 1975. It is heralded for maximum cambium overlap, but is a complex graft. It requires similarly sized diameters for the rootstock and scion. The bark of the rootstock is sliced and peeled back in four flaps, and the hardwood is removed, looking somewhat like a peeled banana.

Awl

Awl grafting takes the least resources and the least time. It is best done by an experienced grafter, as it is possible to accidentally drive the tool too far into the stock, reducing the scion's chance of survival. Awl grafting can be done by using a screwdriver to make a slit in the bark, not penetrating the cambium layer completely. Then inset the wedged scion into the incision.

Veneer

Veneer grafting, or inlay grafting, is a method used for stock larger than 3 cm (1 1⁄8 in) in diameter. The scion is recommended to be about as thick as a pencil. Clefts are made of the same size as the scion on the side of the branch, not on top. The scion end is shaped as a wedge, inserted, and wrapped with tape to the scaffolding branches to give it more strength. May alternately be named a rind graft or a bark graft.

Natural Grafting

A Husband and Wife tree – Natural grafting in blackthorn Prunus spinosa.

Tree branches and more often roots of the same species will sometimes naturally graft; this is called inosculation. The bark of the tree may be stripped away when the roots make physical contact with each other, exposing the vascular cambium and allowing the roots to graft together. A group of trees can share water and mineral nutrients via root grafts, which may be advantageous to weaker trees, and may also form a larger rootmass as an adaptation to promote fire resistance and regeneration as exemplified by the California black oak (Quercus kelloggii). Additionally, grafting may protect the group from wind damages as a result of the increased mechanical stability provided by the grafting. Albino redwoods use root grafting as a form of plant parasitism of normal redwoods.

A problem with root grafts is that they allow transmission of certain pathogens, such as Dutch elm disease. Inosculation also sometimes occurs where two stems on the same tree, shrub or vine make contact with each other. This is common in plants such as strawberries and potato.

Natural grafting is rarely seen in herbaceous plants as those types of plants generally have short-lived roots with little to no secondary growth in the vascular cambium.

Graft Chimera

Occasionally, a so-called "graft hybrid" or more accurately graft chimera can occur where the tissues of the stock continue to grow within the scion. Such a plant can produce flowers and foliage typical of both plants as well as shoots intermediate between the two. The best-known example this is probably +Laburnocytisus 'Adamii', a graft hybrid between Laburnum and Cytisus, which originated in a nursery near Paris, France in 1825. This small tree bears yellow flowers typical of Laburnum anagyroides, purple flowers typical of Cytisus purpureus and curious coppery-pink flowers that show characteristics of both "parents". Many species of cactus can also produce graft chimeras under the right conditions although they are often created unintentionally and such results are often hard to replicate.

Scientific Uses

Grafting has been important in flowering research. Leaves or shoots from plants induced to flower can be grafted onto uninduced plants and transmit a floral stimulus that induces them to flower.

The transmission of plant viruses has been studied using grafting. Virus indexing involves grafting a symptomless plant that is suspected of carrying a virus onto an indicator plant that is very susceptible to the virus.

Grafting can transfer chloroplasts (specialised DNA in plants that can conduct photosynthesis), mitochondrial DNA and the entire cell nucleus containing the genome to potentially make a new species making grafting a form of natural genetic engineering.

White Spruce

White spruce can be grafted with consistent success by using 8–10 cm (3–4 in) scions of current growth on thrifty 4- to 5-year-old rootstock. Before greenhouse grafting, rootstocks should be potted in late spring, allowed to make seasonal growth, then subjected to a period of chilling outdoors, or for about 8 weeks in a cool room at 2 °C.

A method of grafting white spruce of seed-bearing age during the time of seed harvest in the fall was developed by Nienstaedt et al. Scions of white spruce of 2 ages of wood from 30- to 60-year-old trees were collected in the fall and grafted by 3 methods on potted stock to which different day-length treatments had been applied prior to grafting. The grafted stock were given long-day and natural-day treatments. Survival was 70% to 100% and showed effects of rootstock and post-grafting treatments in only a few cases. Photoperiod and temperature treatments after grafting, however, had considerable effect on scion activity and total growth. The best post-grafting treatment was 4 weeks of long-day treatment followed by 2 weeks of short-day treatment, then 8 weeks of chilling, and finally long-day treatment.

Since grafts of white spruce put on relatively little growth in the 2 years after grafting, techniques for accelerating the early growth were studied by Greenwood and others. The cultural regimes used to promote one additional growth cycle in one year involve manipulation of day length and the use of cold storage to satisfy chilling requirements. Greenwood took dormant potted grafts into the greenhouse in early January then gradually raised the temperature during the course of a week until the minimum temperature rose to 15 °C. Photoperiod was increased to 18 hours using incandescent lighting. In this technique, grafts are grown until elongation has been completed, normally by mid-March. Soluble 10-52-10 fertilizer is applied at both ends of the growth cycle and 20-20-20 during the cycle, with irrigation as needed. When growth elongation is complete, day length is reduced to 8 hours using a blackout curtain. Budset follows, and the grafts are held in the greenhouse until mid-May. Grafts are then moved into a cooler at 4 °C for 1000 hours, after which they are moved to a shade frame where they grow normally, with applications of fertilizer and irrigation as in the first cycle. Grafts are moved into cold frames or unheated greenhouse in September until January. Flower induction treatments are begun on grafts that have reached a minimum length of 1.0 m. Repotting from an initial pot size of 4.5 litre to 16 litre containers with a 2:1:1 soil mix of peat moss, loam, and aggregate.

In one of the first accelerated growth experiments, white spruce grafts made in January and February that would normally elongate shortly after grafting, set bud, and remain in that condition until the following spring, were refrigerated for 500, 1000, or 1500 hours beginning in mid-July, and a non-refrigerated control was held in the nursery. After completion of the cold treatment, the grafts were moved into the greenhouse with an 18-hour photoperiod until late October. Height increment was significantly (P 0.01) influenced by cold treatment. Best results were given by the 1000-hour treatment.

The refrigeration (cold treatment) phase was subsequently shown to be effective when applied 2 months earlier with proper handling and use of blackout curtains, which allows the second growth cycle to be completed in time to satisfy dormancy requirements before January.

Herbaceous Grafting

Grafting is often done for non-woody and vegetable plants (tomato, cucumber, eggplant and watermelon). Tomato grafting is very popular in Asia and Europe, and is gaining popularity in the United States. The main advantage of grafting is for disease-resistant rootstocks. Researchers in Japan developed automated processes using grafting robots as early as 1987. Plastic tubing can be used to prevent desiccation and support the healing at the graft/scion interface.

Plant Breeding

Plant breeding is the application of genetic principles to produce plants that are more useful to humans. This is accomplished by selecting plants found to be economically or aesthetically desirable, first by controlling the mating of selected individuals, and then by selecting certain individuals among the progeny. Such processes, repeated over many generations, can change the hereditary makeup and value of a plant population far beyond the natural limits of previously existing populations.

Goals

The plant breeder usually has in mind an ideal plant that combines a maximum number of desirable characteristics. These characteristics may include resistance to diseases and insects; tolerance to heat, soil salinity, or frost; appropriate size, shape, and time to maturity; and many other general and specific traits that contribute to improved adaptation to the environment, ease in growing and handling, greater yield, and better quality. The breeder of horticultural plants must also consider aesthetic appeal. Thus the breeder can rarely focus attention on any one characteristic but must take into account the manifold traits that make the plant more useful in fulfilling the purpose for which it is grown. Plant breeding is an important tool in promoting global food security, and many staple crops have been bred to better withstand extreme weather conditions associated with global warming, such as drought or heat waves.

Increase of Yield

One of the aims of virtually every breeding project is to increase yield. This can often be brought about by selecting obvious morphological variants. One example is the selection of dwarf, early maturing varieties of rice. These dwarf varieties are sturdy and give a greater yield of grain. Furthermore, their early maturity frees the land quickly, often allowing an additional planting of rice or other crop the same year.

Another way of increasing yield is to develop varieties resistant to diseases and insects. In many cases the development of resistant varieties has been the only practical method of pest control. Perhaps the most important feature of resistant varieties is the stabilizing effect they have on production and hence on steady food supplies. Varieties tolerant to drought, heat, or cold provide the same benefit.

Modifications of Range and Constitution

Another common goal of plant breeding is to extend the area of production of a crop species. A good example is the modification of grain sorghum since its introduction to the United States in the 1750s. Of tropical origin, grain sorghum was largely confined to the southern Plains area and the Southwest, but earlier-maturing varieties were developed, and grain sorghum is now an important crop as far north as North Dakota.

Development of crop varieties suitable for mechanized agriculture has become a major goal of plant breeding in recent years. Uniformity of plant characters is very important in mechanized agriculture because field operations are much easier when the individuals of a variety are similar in time of germination, growth rate, size of fruit, and so on. Uniformity in maturity is, of course, essential when crops such as tomatoes and peas are harvested mechanically.

The nutritional quality of plants can be greatly improved by breeding. For example, it is possible to breed varieties of corn (maize) much higher in lysine than previously existing varieties. Breeding high-lysine maize varieties for those areas of the world where maize is the major source of this nutritionally essential amino acid has become a major goal in plant breeding. This "biofortification" of food crops, a term which also includes genetic modification, has been shown to improve nutrition and is especially useful in developing areas where nutritional deficiencies are common and medical infrastructure may be lacking.

In breeding ornamental plants, attention is paid to such factors as longer blooming periods, improved keeping qualities of flowers, general thriftiness, and other features that contribute to usefulness and aesthetic appeal. Novelty itself is often a virtue in ornamentals, and the spectacular, even the bizarre, is often sought.

Evaluation of Plants

The appraisal of the value of plants so that the breeder can decide which individuals should be discarded and which allowed to produce the next generation is a much more difficult task with some traits than with others.

Qualitative Characters

The easiest characters, or traits, to deal with are those involving discontinuous, or qualitative, differences that are governed by one or a few major genes. Many such inherited differences exist, and they frequently have profound effects on plant value and utilization. Examples are starchy versus sugary kernels (characteristic of field and sweet corn, respectively) and determinant versus indeterminant habit of growth in green beans (determinant varieties are adapted to mechanical harvesting). Such differences can be seen easily and evaluated quickly, and the expression of the traits remains the same regardless of the environment in which the plant grows. Traits of this type are termed highly heritable.

Quantitative Characters

In other cases, however, plant traits grade gradually from one extreme to another in a continuous series, and classification into discrete classes is not possible. Such variability is termed quantitative. Many traits of economic importance are of this type; e.g., height, cold and drought tolerance, time to maturity, and, in particular, yield. These traits are governed by many genes, each having a small effect. Although the distinction between the two types of traits is not absolute, it is nevertheless convenient to designate qualitative characters as those involving discrete differences and quantitative characters as those involving a graded series.

Quantitative characters are much more difficult for the breeder to control, for three main reasons: (1) the sheer numbers of the genes involved make hereditary change slow and difficult to assess; (2) the variations of the traits involved are generally detectable only through measurement and exacting statistical analyses; and (3) most of the variations are due to the environment rather than to genetic endowment; for example, the heritability of certain traits is less than 5 percent, meaning that 5 percent of the observed variation is caused by genes and 95 percent is caused by environmental influences.

It follows that carefully designed experiments are required to distinguish plants that are superior because they carry desirable genes from those that are superior because they happen to grow in a favourable site.

Methods of Plant Breeding

Mating Systems

Angiosperm mating systems devolve about the type of pollination, or transferal of pollen from flower to flower. A flower is self-pollinated (a "selfer") if pollen is transferred to it from any flower of the same plant and cross-pollinated (an "outcrosser" or "outbreeder") if the pollen comes from a flower on a different plant. About half of the more important cultivated plants are naturally cross-pollinated, and their reproductive systems include various devices that encourage cross-pollination—e.g., protandry (pollen shed before the ovules are mature, as in the carrot and walnut), dioecy (male and female parts are borne on different plants, as in the date palm, asparagus, and hops), and genetically determined self-incompatibility (inability of pollen to grow on the stigma of the same plant, as in white clover, cabbage, and many other species).

Other plant species, including a high proportion of the most important cultivated plants such as wheat, barley, rice, peas, beans, and tomatoes, are predominantly self-pollinating. There are relatively few reproductive mechanisms that promote self-pollination; the most positive of which is failure of the flowers to open (cleistogamy), as in certain violets. In barley, wheat, and lettuce the pollen is shed before or just as the flowers open, and in the tomato pollination follows opening of the flower, but the stamens form a cone around the stigma. In such species there is always a risk of unwanted cross-pollination.

In controlled breeding procedures it is imperative that pollen from the desired male parent, and no other pollen, reaches the stigma of the female parent. When stamens and pistils occur in the same flower, the anthers must be removed from flowers selected as females before pollen is shed. This is usually done with forceps or scissors. Protection must also be provided from "foreign" pollen. The most common method is to cover the flower with a plastic or paper bag. When the stigma of the female parent becomes receptive, pollen from the desired male parent is transferred to it, often by breaking an anther over the stigma, and the protective bag is replaced. The production of certain hybrids is, therefore, tedious and expensive because it often requires a series of delicate, exacting, and properly timed hand operations. When male and female parts occur in separate flowers, as in corn (maize), controlled breeding is easier.

A cross-pollinated plant, which has two parents, each of which is likely to differ in many genes, produces a diverse population of plants hybrid (heterozygous) for many traits. A self-pollinated plant, which has only one parent, produces a more uniform population of plants pure breeding (homozygous) for many traits. Thus, in contrast to outbreeders, self-breeders are likely to be highly homozygous and hence true breeding for a specified trait.

Breeding Self-pollinated Species

The breeding methods that have proved successful with self-pollinated species are: (1) mass selection; (2) pure-line selection; (3) hybridization, with the segregating generations handled by the pedigree method, the bulk method, or by the backcross method; and (4) development of hybrid varieties.

Mass Selection

In mass selection, seeds are collected from (usually a few dozen to a few hundred) desirable appearing individuals in a population, and the next generation is sown from the stock of mixed seed. This procedure, sometimes referred to as phenotypic selection, is based on how each individual looks. Mass selection has been widely used to improve old "land" varieties—varieties that have been passed down from one generation of farmers to the next over long periods—and is common in horticulture.

An alternative approach that has no doubt been practiced for thousands of years is simply to eliminate undesirable types by destroying them in the field. The results are similar whether superior plants are saved or inferior plants are eliminated: seeds of the better plants become the planting stock for the next season.

A modern refinement of mass selection is to harvest the best plants separately and to grow and compare their progenies. The poorer progenies are destroyed and the seeds of the remainder are harvested. It should be noted that selection is now based not solely on the appearance of the parent plants but also on the appearance and performance of their progeny. Progeny selection is usually more effective than phenotypic selection when dealing with quantitative characters of low heritability. It should be noted, however, that progeny testing requires an extra generation; hence gain per cycle of selection must be double that of simple phenotypic selection to achieve the same rate of gain per unit time.

Mass selection, with or without progeny test, is perhaps the simplest and least expensive of plant-breeding procedures. It finds wide use in the breeding of certain forage species, which are not important enough economically to justify more detailed attention.

Pure-line Selection

Pure-line selection generally involves three more or less distinct steps: (1) numerous superior appearing plants are selected from a genetically variable population; (2) progenies of the individual plant selections are grown and evaluated by simple observation, frequently over a period of several years; and (3) when selection can no longer be made on the basis of observation alone, extensive trials are undertaken, involving careful measurements to determine whether the remaining selections are superior in yielding ability and other aspects of performance. Any progeny superior to an existing variety is then released as a new "pure-line" variety. Much of the success of this method during the early 1900s depended on the existence of genetically variable land varieties that were waiting to be exploited. They provided a rich source of superior pure-line varieties, some of which are still represented among commercial varieties. In recent years the pure-line method as outlined above has decreased in importance in the breeding of major cultivated species; however, the method is still widely used with the less important species that have not yet been heavily selected.

A variation of the pure-line selection method that dates back centuries is the selection of single-chance variants, mutations or "sports" in the original variety. A very large number of varieties that differ from the original strain in characteristics such as colour, lack of thorns or barbs, dwarfness, and disease resistance have originated in this fashion.

Hybridization

During the 20th century planned hybridization between carefully selected parents has become dominant in the breeding of self-pollinated species. The object of hybridization is to combine desirable genes found in two or more different varieties and to produce pure-breeding progeny superior in many respects to the parental types.

Genes, however, are always in the company of other genes in a collection called a genotype. The plant breeder's problem is largely one of efficiently managing the enormous numbers of genotypes that occur in the generations following hybridization. As an example of the power of hybridization in creating variability, a cross between hypothetical wheat varieties differing by only 21 genes is capable of producing more than 10,000,000,000 different genotypes in the second generation. While the great majority of these second generation genotypes are hybrid (heterozygous) for one or more traits, it is statistically possible that 2,097,152 different pure-breeding (homozygous) genotypes can occur, each potentially a new pure-line variety. These numbers illustrate the importance of efficient techniques in managing hybrid populations, for which purpose the pedigree procedure is most widely used.

Pedigree breeding starts with the crossing of two genotypes, each of which have one or more desirable characters lacked by the other. If the two original parents do not provide all of the desired characters, a third parent can be included by crossing it to one of the hybrid progeny of the first generation (F1). In the pedigree method superior types are selected in successive generations, and a record is maintained of parent–progeny relationships.

The F2 generation (progeny of the crossing of two F1 individuals) affords the first opportunity for selection in pedigree programs. In this generation the emphasis is on the elimination of individuals carrying undesirable major genes. In the succeeding generations the hybrid condition gives way to pure breeding as a result of natural self-pollination, and families derived from different F2 plants begin to display their unique character. Usually one or two superior plants are selected within each superior family in these generations. By the F5 generation the pure-breeding condition (homozygosity) is extensive, and emphasis shifts almost entirely to selection between families. The pedigree record is useful in making these eliminations. At this stage each selected family is usually harvested in mass to obtain the larger amounts of seed needed to evaluate families for quantitative characters. This evaluation is usually carried out in plots grown under conditions that simulate commercial planting practice as closely as possible. When the number of families has been reduced to manageable proportions by visual selection, usually by the F7 or F8 generation, precise evaluation for performance and quality begins. The final evaluation of promising strains involves (1) observation, usually in a number of years and locations, to detect weaknesses that may not have appeared previously; (2) precise yield testing; and (3) quality testing. Many plant breeders test for five years at five representative locations before releasing a new variety for commercial production.

The bulk-population method of breeding differs from the pedigree method primarily in the handling of generations following hybridization. The F2 generation is sown at normal commercial planting rates in a large plot. At maturity the crop is harvested in mass, and the seeds are used to establish the next generation in a similar plot. No record of ancestry is kept. During the period of bulk propagation natural selection tends to eliminate plants having poor survival value. Two types of artificial selection also are often applied: (1) destruction of plants that carry undesirable major genes and (2) mass techniques such as harvesting when only part of the seeds are mature to select for early maturing plants or the use of screens to select for increased seed size. Single plant selections are then made and evaluated in the same way as in the pedigree method of breeding. The chief advantage of the bulk population method is that it allows the breeder to handle very large numbers of individuals inexpensively.

Often an outstanding variety can be improved by transferring to it some specific desirable character that it lacks. This can be accomplished by first crossing a plant of the superior variety to a plant of the donor variety, which carries the trait in question, and then mating the progeny back to a plant having the genotype of the superior parent. This process is called backcrossing. After five or six backcrosses the progeny will be hybrid for the character being transferred but like the superior parent for all other genes. Selfing the last backcross generation, coupled with selection, will give some progeny pure breeding for the genes being transferred. The advantages of the backcross method are its rapidity, the small number of plants required, and the predictability of the outcome. A serious disadvantage is that the procedure diminishes the occurrence of chance combinations of genes, which sometimes leads to striking improvements in performance.

Hybrid Varieties

The development of hybrid varieties differs from hybridization in that no attempt is made to produce a pure-breeding population; only the F1 hybrid plants are sought. The F1 hybrid of crosses between different genotypes is often much more vigorous than its parents. This hybrid vigour, or heterosis, can be manifested in many ways, including increased rate of growth, greater uniformity, earlier flowering, and increased yield, the last being of greatest importance in agriculture.

By far the greatest development of hybrid varieties has been in corn (maize), primarily because its male flowers (tassels) and female flowers (incipient ears) are separate and easy to handle, thus proving economical for the production of hybrid seed. The production of hand-produced F1 hybrid seed of other plants, including ornamental flowers, has been economical only because greenhouse growers and home gardeners have been willing to pay high prices for hybrid seed.

Recently, however, a built-in cellular system of pollination control has made hybrid varieties possible in a wide range of plants, including many that are self-pollinating, such as sorghums. This system, called cytoplasmic male sterility, or cytosterility, prevents normal maturation or function of the male sex organs (stamens) and results in defective pollen or none at all. It obviates the need for removing the stamens either by hand or by machine. Cytosterility depends on the interaction between male sterile genes (R + r) and factors found in the cytoplasm of the female sex cell. The genes are derived from each parent in the normal Mendelian fashion, but the cytoplasm (and its factors) is provided by the egg only; therefore, the inheritance of cytosterility is determined by the female parent. All plants with fertile cytoplasm produce viable pollen, as do plants with sterile cytoplasm but at least one R gene; plants with sterile cytoplasm and two r genes are male sterile (produce defective pollen).

The production of F1 hybrid seed between two strains is accomplished by interplanting a sterile version of one strain (say A) in an isolated field with a fertile version of another strain (B). Since strain A produces no viable pollen, it will be pollinated by strain B, and all seeds produced on strain A plants must therefore be F1 hybrids between the strains. The F1 hybrid seeds are then planted to produce the commercial crop. Much of the breeder's work in this process is in developing the pure-breeding sterile and fertile strains to begin the hybrid seed production.

Breeding Cross-pollinated Species

The most important methods of breeding cross-pollinated species are (1) mass selection; (2) development of hybrid varieties; and (3) development of synthetic varieties. Since cross-pollinated species are naturally hybrid (heterozygous) for many traits and lose vigour as they become purebred (homozygous), a goal of each of these breeding methods is to preserve or restore heterozygosity.

Mass Selection

Mass selection in cross-pollinated species takes the same form as in self-pollinated species; i.e., a large number of superior appearing plants are selected and harvested in bulk and the seed used to produce the next generation. Mass selection has proved to be very effective in improving qualitative characters, and, applied over many generations, it is also capable of improving quantitative characters, including yield, despite the low heritability of such characters. Mass selection has long been a major method of breeding cross-pollinated species, especially in the economically less important species.

Development of Hybrid Varieties

The outstanding example of the exploitation of hybrid vigour through the use of F1 hybrid varieties has been with corn (maize). The production of a hybrid corn variety involves three steps: (1) the selection of superior plants; (2) selfing for several generations to produce a series of inbred lines, which although different from each other are each pure-breeding and highly uniform; and (3) crossing selected inbred lines. During the inbreeding process the vigour of the lines decreases drastically, usually to less than half that of field-pollinated varieties. Vigour is restored, however, when any two unrelated inbred lines are crossed, and in some cases the F1 hybrids between inbred lines are much superior to open-pollinated varieties. An important consequence of the homozygosity of the inbred lines is that the hybrid between any two inbreds will always be the same. Once the inbreds that give the best hybrids have been identified, any desired amount of hybrid seed can be produced.

Pollination in corn (maize) is by wind, which blows pollen from the tassels to the styles (silks) that protrude from the tops of the ears. Thus controlled cross-pollination on a field scale can be accomplished economically by interplanting two or three rows of the seed parent inbred with one row of the pollinator inbred and detasselling the former before it sheds pollen. In practice most hybrid corn is produced from "double crosses," in which four inbred lines are first crossed in pairs (A × B and C × D) and then the two F1 hybrids are crossed again (A × B) × (C × D). The double-cross procedure has the advantage that the commercial F1 seed is produced on the highly productive single cross A × B rather than on a poor-yielding inbred, thus reducing seed costs. In recent years cytoplasmic male sterility, described earlier, has been used to eliminate detasselling of the seed parent, thus providing further economies in producing hybrid seed.

Much of the hybrid vigour exhibited by F1 hybrid varieties is lost in the next generation. Consequently, seed from hybrid varieties is not used for planting stock but the farmer purchases new seed each year from seed companies.

Perhaps no other development in the biological sciences has had greater impact on increasing the quantity of food supplies available to the world's population than has the development of hybrid corn (maize). Hybrid varieties in other crops, made possible through the use of male sterility, have also been dramatically successful and it seems likely that use of hybrid varieties will continue to expand in the future.

Development of Synthetic Varieties

A synthetic variety is developed by intercrossing a number of genotypes of known superior combining ability—i.e., genotypes that are known to give superior hybrid performance when crossed in all combinations. (By contrast, a variety developed by mass selection is made up of genotypes bulked together without having undergone preliminary testing to determine their performance in hybrid combination.) Synthetic varieties are known for their hybrid vigour and for their ability to produce usable seed for succeeding seasons. Because of these advantages, synthetic varieties have become increasingly favoured in the growing of many species, such as the forage crops, in which expense prohibits the development or use of hybrid varieties.

Distribution and Maintenance of New Varieties

The benefits of superior new varieties obviously cannot be realized until sufficient seed has been produced to permit commercial production. Although the primary function of the plant breeder is to develop new varieties, he usually also carries out an initial small-scale seed increase. Seed thus produced is called breeders seed. The next stage is the multiplication of breeders seed to produce foundation seed. Production of foundation seed is usually carried out by seed associations or institutes, whose work is regulated by government agencies. The third step is the production of certified seed, the progeny of foundation seed, produced on a large scale by specialized seed growers for general sale to farmers and gardeners. Certified seed must be produced and handled in such a way as to meet the standards set by the certifying agency (usually a seed association). Seed associations are also usually responsible for maintaining the purity of new varieties once they have been released for commercial production.

The distribution of new varieties developed by commercial plant-breeding companies is often through seed associations, but many reputable companies market their products without following the official certification process. In some countries, particularly in Europe, new varieties can be patented for periods up to 15 years or more, during which time the breeder has an exclusive right to reproduce and sell the variety.

References

- Horti-Landscaping-nursery%20management, horticulture: tnau.ac.in, Retrieved 29 June, 2019
- Boyle, John (2017-03-16). "Answer Man: What's with the tree hacking? Stop the topping!". Asheville Citizen-Times. Retrieved 2017-03-16
- Propagation-of-plants, science: britannica.com, Retrieved 13 March, 2019
- Aina, O; Quesenberry, K.; Gallo, M (2012). "In vitro induction of tetraploids in Arachis paraguariensis". Plant Cell, Tissue and Organ Culture (PCTOC). 111(2): 231–238. Doi:10.1007/s11240-012-0191-0
- Nelson, Alexander (2007). Principles of Agricultural Botany. United Kingdom: Read Books. P. 101. ISBN 978-1-4067-4662-4
- Germination: biologydictionary.net, Retrieved 10 July, 2019
- Plant-breeding, plant-breeding: britannica.com, Retrieved 23 February, 2019

Organic Horticulture and Gardening 4

- Organic Gardening
- Organic Production of Horticultural Crops
- Organic Fruit Production
- Organic Food

The practice of growing vegetables, fruits, flowers and ornamental plants using the essential principles of organic agriculture is referred to as organic horticulture. Some of the major sub-domains of organic horticulture are organic gardening, organic production of horticultural crops and organic fruit production. The topics elaborated in this chapter will help in gaining a better perspective about organic horticulture.

Organic horticulture is the science and art of growing fruits, vegetables, flowers, or ornamental plants by following the essential principles of organic agriculture in soil building and conservation, pest management, and heirloom variety preservation, Organic horticulture (or organic gardening) is based on knowledge and techniques gathered over thousands of years. In general terms, organic horticulture involves natural processes, often taking place over extended periods of time, and a sustainable, holistic approach - while chemical-based horticulture focuses on immediate, isolated effects and reductionist strategies.

Organic practices arise from the understanding that all organisms in nature are interdependent, and in order to have healthy plants we must foster the health of their entire ecosystem. These practices go beyond integrated pest management, beyond the use of so-called organic fertilizers and pesticides. They acknowledge the concept of intrinsic health, and seek to create environments that cater to the well-being of all their inhabitants.

Organic Gardening Systems

There are a number of formal organic gardening and farming systems that prescribe specific techniques. They tend to be more specific than, and fit within, general organic standards. Biodynamic farming is an approach based on the esoteric teachings of Rudolf Steiner. The Japanese farmer and writer Masanobu Fukuoka invented a no-till system for small-scale grain production that he called Natural Farming. French intensive and biointensive methods and SPIN Farming (Small Plot

INtensive) are all small scale gardening techniques. A garden is more than just a means of providing food, it is a model of what is possible in a community - everyone could have a garden of some kind (container, growing box, raised bed) and produce healthy, nutritious organic food, a farmers market, a place to pass on gardening experience, and a sharing of bounty, promoting a more sustainable way of living that would encourage their local economy. A simple 4′ × 8′ (32 square feet) raised bed garden based on the principles of bio-intensive planting and square foot gardening uses fewer nutrients and less water, and could keep a family, or community, supplied with an abundance of healthy, nutritious organic greens, while promoting a more sustainable way of living.

Other methods can also be used to supplement an existing garden. Methods such as composting, or vermicomposting. These practices are ways of recycling organic matter into some of the best organic fertilizers and soil conditioner. Vermicompost is especially easy. The byproduct is also an excellent source of nutrients for an organic garden.

Soil

"Soil is the result of interaction between three equal partners: the a-biotic components of the soil, living organisms and environmental components (temperature, water, air). Each of these brings different assets to the partnership, and each makes a unique value contribution to the whole. The synthesis of all contributions - soil - is greater than the sum of its parts. Soil must be understood and managed from this holistic perspective."

"The primary a-biotic component of soil is rock which, as it becomes reduced to smaller particles, contributes the secondary components: sand, silt, clay and mineral nutrients." The "combination of particle size, shape and arrangement affects the soil's compactability, air supply and water holding capacity."

Living organisms make the following contributions to the soil partnership: conversion of carbon and nitrogen gas into solid compounds, modification of inherent soil fertility through biological transmutation, conversion of mineral soil nutrients into organic forms, creation of multi-level food production and storage systems, and creation of habitat through structural improvement of the soil. Environmental components of soil include air, water and temperature. "The environmental components literally become incorporated into living organisms, and into the soil ecosystem. They are not just 'factors' or 'influences', they contribute essential building blocks and energy to fuel the processes. They are vital and active participants."

Soil is an ecosystem "created through the proportionate availability and interaction of all three soil partners. Soil is the habitat living organisms have synergistically created for themselves. Our job is to support them, not to put them out of work." Managing the soil is very important. If your garden is healthy then insects will not attack the plants. Insects only attack plants that are unhealthy. To keep your garden healthy give it organic matter and humus to survive. The most important thing is to give your garden lots of attention and your energy.

Nutrient Management

Principally, animal manures, compost, green manures, bio-fertilizers, mixed organic fertilizers are used in organic farming. Nitrogen is provided by legume crops having nitrogen fixing symbiotic bacteria and by soil habiting non-symbiotic bacteria. Enrichment of phosphorus in soil is done by incorporation of rock-phosphate, VAM (Vesicular Arbuscular Mycorrhiza - solublises phosphorus for greater availability to the plants) and VAM treated compost. Potassium is provided by wood ash, sea weeds, tobacco stem; used alone or in combination with others. Beside it permanent mulching layer reduces the potassium leaching.

Apply lime 2-3 months before planting to correct soil acidity.

Pest Control Approaches

Differing approaches to pest control are equally notable. In chemical horticulture, a specific insecticide may be applied to quickly kill off a particular insect pest. Chemical controls can dramatically reduce pest populations in the short term, yet by unavoidably killing (or starving) natural control insects and animals, cause an increase in the pest population in the long term, thereby creating an ever increasing problem. Repeated use of insecticides and herbicides also encourages rapid natural selection of resistant insects, plants and other organisms, necessitating increased use, or requiring new, more powerful controls.

In contrast, organic horticulture tends to tolerate some pest populations while taking the long view. Organic pest control requires a thorough understanding of pest life cycles and interactions, and involves the cumulative effect of many techniques, including:

- Allowing for an acceptable level of pest damage.
- Encouraging predatory beneficial insects to flourish and eat pests.
- Encouraging beneficial microorganisms.
- Careful plant selection, choosing disease-resistant varieties.
- Planting companion crops that discourage or divert pests.
- Using row covers to protect crop plants during pest migration periods.
- Rotating crops to different locations from year to year to interrupt pest reproduction cycles.
- Using insect traps to monitor and control insect populations.

Each of these techniques also provides other benefits, such as soil protection and improvement, fertilization, pollination, water conservation and season extension. These benefits are both complementary and cumulative in overall effect on site health. Organic pest control and biological pest control can be used as part of integrated pest management (IPM). However, IPM can include the use of chemical pesticides that are not part of organic or biological techniques.

Pest, Disease and Weed Management

Suitable crop rotations, green manuring, use of balanced fertilizers, proper care during nursery, mulching etc. play important role to protect the crop from insect pest and disease. The following practices are recommanded:

- Use trichoderma for seed treatment.
- Use resistant varieties.

- Use disease free planting material.
- Use mulching for those vegetables whose fruits touches the soil.
- Clean the crop residue.
- Summer fallowing and flooding.
- Use spray Bt, NPV, Beauveria.
- Use insect traps- such as pheromone trap, light trap, yellow trap, sticky trap.
- Use botanicals like neem, garlic etc.
- Use trap crops to misguide the insect and protect the main crop.

Organic Gardening

Organic gardening is essentially gardening without using synthetic products like fertilizers and pesticides. It involves the use of only natural products to grow plants in your garden. Organic gardening replenishes the natural resources as it uses them. In organic gardening, you consider your plants as part of the larger natural system that begins with the soil and includes water supply, the wildlife; insects and people. Everyone wants the food we serve to our families as well as our environment to be safe and healthy. A good organic gardener strives to ensure that his or her activities are in harmony with the natural ecosystem and aims at minimizing exploitation as well as replenishing all the resources consumed by his or her garden.

Gardeners and people who have come across the word organic gardening probably usually desire to know what it means. Organic gardening is a terminology that simply refers to growing of plants, vegetables, and fruits in the best natural way without the use of pesticides or synthetic chemical fertilizers. Even so, organic gardening is more than simply avoiding the use pesticides and synthetic fertilizers.

Gardening organically encompasses supporting the health of the entire gardening system naturally. It means working in harmony with the natural systems including the soil, water supply, people, and even insects with an ultimate aim of minimizing destruction to living and non-living things in the natural environment while constantly replenishing any resources utilized during gardening.

Organic gardening fundamentals requires cultivation emphasized on creating an ecosystem that nourishes and sustains soil microbes and plants, while also benefiting insects rather than just putting seeds in the ground and letting them grow.

To begin with, there are three major areas to concentrate on to maintain the objectives of organic gardening. These includes: soil management which is dealt with by using organic fertilizer; weed management which is managed by manual labor and use of organic ground coverings; and lastly pest control which is dealt with by promoting beneficial insects and companion planting. These are the top key strategies for becoming an organic gardener. Proper knowledge is essential in organic gardening and requires simple fundamentals to get reliable results.

To sum up, organic gardening is as simple as relying on intermingling plant types and varieties, use of companion planting, dense planting in order for some plants to offer companion to vulnerable plants, and supporting natural systems to minimize spread of pests and diseases.

Organic Gardening magazine editor, Therese Ciesinski, describes organic gardening as:

> "Organic gardening is more than simply avoiding synthetic pesticides and fertilizers. It is about observing nature's processes, and emulating them in your garden as best you can. And the most important way to do that is to understand the makeup of your soil and to give it what it needs. If anything could be called a 'rule' in organic gardening, it's this: feed the soil, not the plant."

Step by Step Process to Start an Organic Garden

The following is a step by step process on how to start an organic garden:

1. Prepare the Soil: The soil is the most important thing or resource when it comes to organic gardening. It is achieved through continuous addition of organic matter to the soil by using locally available resources in every possible aspect. If you want your plants to be healthy, you will need to thoroughly prepare the soil on which they will grow on.

Just like human beings, plants require food and the food in this case comes from the soil. Therefore, you need to ensure that your plants get plenty of fresh nutrients. Proper soil conditioning will give your plants all the nutrients they need.

Chemical soil treatments not only destroy the soil composition but they also harm the important microorganisms, worms, and bacteria in the soil. To begin, you will have to test the soil PH. You can do this by buying a home testing kit or simply collect some soil samples and send them to the local agricultural extension office for proper testing and analysis.

2. Make Good Compost: As you wait for soil sample results, you can make the compost. Compost

helps in providing plants with nutrients, helps conserve water, helps in the reduction of weeds and helps in keeping food as well as yard waste out of landfills.

Compost can be obtained or made from locally available resources such as leaves, grass trimmings, yard garbage/remains, and kitchen waste. Alternatively, compost is readily available for purchase from mulch suppliers or organic garden centers. You can use these steps in making compost.

- Measure out space that is at least three square feet.
- Get a pile of natural dead plants or leaves.
- Add alternating layers of leaves, garden trimmings (carbon) and nitrogen (green) materials, for instance, kitchen leftovers and manure. Put a layer or separate them with a layer of soil.
- Cover the pile with about 4-6 inches of soil.
- Turn the pile every time a new layer is added to the mixture. During this process, ensure that water is added to keep the mixture moist to enhance microorganism activity.
- The compost should not smell but in case it does, add some more dry leaves, saw dust or straw and then turn it regularly. Do this for about four weeks or a month and you will have good compost needed for your organic garden.

3. Prepare Your Garden: Once you are waiting for the compost to be ready, the next step is to prepare your garden. After receiving a go ahead from your local agricultural extension officer concerning the right soil type, it is now time to prepare the garden area. Using available gardening tools you can carefully prepare your garden. However, it is important to ensure that you do not completely destroy the soil.

4. Choose the Right Plants: Once you are through with preparing your garden, the next step is to select the right plants for your garden. Soil sampling and testing will come in handy at this stage. It is important to choose plants that will thrive well in specific micro-conditions of your soil type. Carefully choose plants that will thrive well in different spots in your garden in terms of moisture, light, and drainage as well as the soil quality. Remember the healthier your plants are, the more and more resistant your crops will be to attackers.

Another mechanism for growing organically is to select plants suited to the garden. Crops well adapted to the gardens climate and conditions are better able to grow with minimal input. Also, growing crops well adapted to the site ensure greater natural defenses. Meaning, little attention and input is required for boosting crop productivity.

When purchasing seedlings, ensure that you go for plants that are raised without synthetic chemicals or pesticides. Your local farmers market is a good place to do your purchases. You will not only find a variety of plants but plant varieties that are best suited for your local area. Carefully select plants that look healthy and are without overcrowded roots.

5. Plant the Crops in Beds: When planting your crops ensure that you plant them in wide beds. Planting them in beds prevents you from walking on them and destroying the soil surface when harvesting or when cutting the flowers. Additionally, grouping crops helps in reducing weeding,

wastage of water and makes it easier for you to apply compost. It also enables the plants to utilize the available nutrients and water. Ensure that there is adequate space between the rows. This helps in promoting air circulation which helps in repelling fungal attacks.

6. Water the Crops: Once you have planted your crops, the next step is to water them. It is good to water the plants immediately after planting them to give them the much-needed water to enable them to continue growing. You can also water them every morning. It is recommended to water your pants in the morning because there are no strong winds, mornings are cool and the water lost as a result of evaporation is immensely reduced. Experts recommend considerable, infrequent watering for plants that are already established.

7. Weeding: According to CSU, Weeds reduce crop yield by competing for water, light, soil nutrients, and space. In agricultural crops, weeds can reduce crop quality by contaminating the commodity. They can serve as hosts for diseases or provide shelter for insects to overwinter. It may be a hard work to pull out weeds by hand but then its a good exercise that helps you to get some fresh air.

8. Provide Nutrients to Your Plants: When you do organic gardening, you need to look out for eco-friendly ways to protect your plants from toxic pesticides and fertilizers. So, you need to make sure that your plants get enough light, nutrients and moisture that help them to grow better. Also, a diverse garden would help to prevent pests, by limiting the amount of one type of plant and gives boost to biodiversity.

Instead of using chemical pesticides, organic gardening emphasizes on promoting beneficial insects and companion planting. Organic farmers need not to eliminate insects and diseases by using chemicals. Rather, pest control is done through keeping pests and diseases below damaging levels. One of the major mechanisms is by promoting beneficial insects and pest predators such as bats, birds, lizards, toads, and spiders.

The key to succeeding in this area is to grow a wide variety of companion crops that supports the ecological niche of these species. Avoiding the use of synthetic pesticides as well ensures their survival. Nipping off infected/infested leaves or buds, uprooting infected crops, crop rotation, and handpicking insect pests and eggs are excellent methods for controlling pest populations. Maintaining garden cleanliness is another effective tool for organic pest control.

Natural sprays or pesticides can equally be used in addition to the cultural pest control methods. They are readily available from organic garden centers and their products contain the bacterium Bacillus, neem oil, and minerals like copper. Their ability to break down quickly has promoted their wide usage in place of the synthetic chemical pesticides. Besides, there are some vegetable/fruit pests and diseases that are beyond natural and organic control, thereby requiring the use of natural sprays.

9. Use of Organic Fertilizers: As much as organic matter and compost will improve water and nutrient retention in the soil, the supply of all the required nutrients for healthy and productive growth is limited. Besides compost, organic gardening requires additional fertilizers drawn from natural sources such as: plant products like wood ash; natural deposits for example rock phosphate; and animal byproducts as well as manures.

Agricultural lime is another natural product frequently added to soil to improve its quality. It is produced from naturally occurring limestone and is added to the soil to optimize pH if the soil is too acidic. Soil pH levels vary from one locality to another. Local extension offices usually provide guidance on soil pH level testing within their jurisdiction areas. However, most soils do not need additional liming.

10. Taking Care of Your Organic Garden: There are important maintenance tips for your organic garden if you want to maintain or harvest healthy plants.

- Mulching,
- Watering in the mornings,
- Use of compost,
- Use of natural manure especially from animals that do not eat meat,
- Weeding at the right time,
- Prune regularly to allow for proper aeration and effective use of nutrients including light.

Various Benefits of Organic Gardening

Organic gardening focuses on the use of only natural products to grow plants. This means that there is no use of pesticides. Therefore, the crops obtained from this type of gardening are free of pesticides and other chemicals. You will live a healthy life without worrying about consuming chemicals through consuming crops.

Organic gardening promises good health because the produce is free from toxic ingredients and other synthetically enhanced chemicals. The fruits and vegetables grown in the organic garden do not have the chemical residues which enter the body when eaten. Organic veggies are also proven to contain higher mineral and vitamin contents compared to those grown using pesticides, herbicides, and chemical fertilizers.

Furthermore, complementary medicine professionals affirm that there are high nutrient concentrations such as Vitamins C and D in organic food products. Organic gardening also adds an extra benefit of exercising the body, especially from manual labor including planting, weeding, and harvesting. Thus, gardening can be an enjoyable and meaningful way to boost physical exercise. The outdoor environment also offers a refreshing means of connecting with nature, sunshine and fresh air, acting as a stress reliever.

Helps in Conserving the Environment

Use of chemicals on crops is one way in which we pollute the environment. The chemicals sprayed or applied on the crops seethe through the soil into the water. This puts the microorganisms at risk. When spraying the crops, the wind carries away the chemicals into the atmosphere, and this is air pollution. Therefore, embracing organic gardening is one of the best ways of maintaining a healthy environment.

Choosing organic gardening massively aids in environmental preservation. Growing vegetables and fruits in a natural way will not only ensure healthy produce but also promote a friendly and toxicity free environment. Organic gardening is the surest way of preserving a healthy and green environment by ensuring ecological balance and minimal disturbance to the natural environment.

It ensures birds, small animals, and beneficial insects are free from chemical harm. Organic matter used to prepare the soil also helps to improve the soil quality. Therefore, organic gardening offers the most beneficial outcome for good environmental health. Putting an end to chemical fertilizers and pesticides that leach into the ground and find way into water supply is only a possibility through organic gardening.

Reduces Greenhouse Gas Emissions

Crops that contain chemicals release these chemicals into the atmosphere in the form of greenhouse gasses. These gasses combined with other impurities in the air pose a risk to the air we breathe in. Additionally, these gasses also contribute immensely to global warming, and we all know the negative effects of global warming.

Better Taste

Organic vegetables and fruits contain scent and taste which are simply better due to their natural growth compared to those grown commercially. To a very large extent, veggies and fruits grown commercially cannot bear or beat the natural flavors of those that are grown organically. Fresh vegetables and fruits from the garden have always tasted better and have natural flavor.

Saves Money

A tremendous way of saving money is by growing your own organic vegetable garden. Saving money is something that every person wants to do but it can only be realized by undertaking small initiatives like growing own veggies and fruits. Through organic gardening, one can save up to 50% of the money used to buy fruits and veggies at supermarkets as well as other perishable stores.

To conclude, organic gardening is beneficial to the people and the environment. However, learning how to start an organic is a very important step in ensuring that you conserve the environment and that you grow your own delicious fresh produce. To give your crops the needed nourishment, you can use natural fertilizers, for example, seaweed extracts, fish emulsion or manures obtained from animal droppings especially cow and chicken droppings that is readily available or bought from your local garden center.

Organic Production of Horticultural Crops

Producing Organic Horticultural Crops

Organic production involves a combination of management techniques to maintain soil quality and fertility while controlling weeds, pests and diseases. Crop rotation plays a big role in achieving these goals.

There are several accredited organic certification organizations in Saskatchewan. Their standards vary slightly, partly due to different interpretations of "restricted use" products. Organic producers must pay attention to the requirements of the agency they choose to certify their production.

Consumers often interpret "certified organic" produce as merely pesticide free, but this is not always the case. Approved organic pesticides may have been used in some cases. Organic crops must always be produced according to accepted guidelines of the organic standard being followed, including soil management practices.

Vegetable Crops

While home gardeners can grow the majority of vegetables "organically," the list of crops suitable for organic production is limited for commercial production. Markets for organic produce have increased in the last decade, but some vegetables are considered too high risk for commercial organic production, particularly cabbage and related vegetables. Pest control for these crops is very difficult and is compounded if the crops are located in regions where canola and mustard are being grown. If they are to be attempted, these crops need to be well isolated from crops that share the same pests.

Carrots

Carrots are considered to be one of the easiest crops to grow organically due to the limited number of pests. However, they should not be planted near alfalfa or other hay and winter cereals, due to the incidence of leafhoppers in these crops, which can spread aster yellows disease to carrots. The disease causes stunting or discoloration of carrot foliage, and extremely hairy roots that are unmarketable. As carrots are slow to germinate and are poor weed competitors when young, weed control is generally the greatest challenge in growing carrots.

Pumpkins and Squash

Pumpkins and the various types of summer and winter squash, including zucchini, are also potential crops for organic production. They are not prone to many insect pests or diseases in Saskatchewan. Selecting early maturing winter squash is important due to the short growing season in Saskatchewan.

Potatoes

While not highly recommended for organic production in Saskatchewan, a few producers have successfully grown potatoes organically. Disease control, particularly for blight, is of major concern. Insects such as Colorado potato beetles, flea beetles and wireworms can also be major problems. This crop is best grown well isolated from other potato fields and with long rotations. Careful cultivar selection may assist in disease control.

Sweet Corn

Sweet corn is a popular item at farmers' markets, roadside stands, supermarkets and market gardens. One of the greatest challenges in growing sweet corn is providing it with sufficient nitrogen to meet its very high requirements. Growers also need to look for untreated seed. Corn seed maggot, corn earworm and European corn borer are potential pests.

Onions

Organic production of onions is very risky, due to the high likelihood the crop will be infested with onion maggots. Without the use of conventional pesticides, it may be very difficult to produce a profitable, marketable crop. If onions are to be planted, it is suggested to plant the crop about 1.6 km (1 mile) from previous onion cropland. Although they prefer onions, onion maggots can also infest garlic, leeks, chives and shallots. Onion thrips are another common problem. They will migrate from alfalfa and cereal crops at harvest time. Heavy irrigation is known to reduce thrips populations.

Cole Crops (Cabbage Family)

Vegetables in this group are not recommended for organic production in Saskatchewan, due to the high incidence of pests. This family includes: cabbage, cauliflower, broccoli, Brussels sprouts, kohlrabi, kale, radish, rutabaga and turnip. They are very prone to imported cabbageworm, diamondback moth larvae, root maggots, red turnip beetle, flea beetles, cabbage thrips and wireworms. The incidence will be higher in regions where canola and mustard crops are being grown. Of the group, rutabagas and turnips are particularly difficult to protect organically from root maggots.

Other Vegetables

Many other vegetables are somewhat intermediate in their ease of cultivation by organic methods. Some of them are listed here, along with potential problems:

- Asparagus - this perennial crop is not widely grown in Saskatchewan. The early spears can be a target for overwintering onion thrips. Asparagus beetles, exclusive to this crop, can also feed on spears and make them unmarketable. Crown and root rot may be managed to some extent by cultural practices.

- Beans and Peas - untreated seed must be used, making them more prone to seed and/or root rot. Powdery mildew is a common problem in peas, particularly late in the season.

- Beets - if marketed with tops attached, beet leafminer control will be necessary. Row covers may be considered. Adjacent weeds such as lamb's-quarters are also hosts for leafminer. Rotations are helpful in minimizing problems with leaf and root diseases. Swiss chard and spinach share some of the same pests and diseases.

- Celery - this crop needs an ample water supply at all times to keep growth succulent. Water stress will cause tough, strong flavoured stems. Tarnished plant bug can cause serious damage to this crop; thus it should not be grown near alfalfa fields.

- Cucumbers - are generally easy to grow, but need a consistent water supply, especially when fruit is developing. Cucumber beetles can spread bacterial wilt disease. Transplanting and the use of row covers early in the season will provide earlier harvests.

- Lettuce - leaf lettuce is the easiest to produce but highly perishable when harvested. Romaine and head lettuce are later maturing. Lettuce should be marketed as soon as possible after harvest. Tipburn resistant varieties should be grown. Lettuce is prone to various diseases including Sclerotinia (white mould). Using raised beds and rotating fields with grain crops may be helpful. Aster yellows disease, spread by leafhoppers, may also be a potential problem, but the incidence is quite variable from year to year.
- Melons - both muskmelon (often called cantaloupe) and watermelon are long-season crops that need warm weather to produce well. Production in tunnels, on plastic mulch or season-extending greenhouse-type structures has given earlier yields. These crops are best grown on sandy land that warms up easily.
- Peppers - by both organic and conventional methods, unless grown in a protected structure, peppers are somewhat of a gamble in Saskatchewan. They prefer a warm, long growing season, although excessively hot summer temperatures can cause flower bud drop.
- Spinach - beet leafminers are a major problem in this crop, so it may not be profitable unless it can be grown in isolation or under row covers.
- Tomatoes - are susceptible to early and late blight, cutworms, and sometimes to Colorado potato beetles. In Saskatchewan, early maturing and cold-tolerant cultivars are required unless the plants are to be grown in greenhouses.

Herb Production

Culinary Herbs

A wide range of herbs can be grown outdoors or in greenhouses. Some herbs such as sweet basil and rosemary are very adaptable to greenhouse production. Organic garlic is in demand in some areas, but production may not be practical on a large scale because garlic must be kept weed free and can be affected by some of the same pests as onions. Growers may find the garlic seed market to be more lucrative than the culinary market.

Most culinary herb production needs to take place near urban centres where farmers' markets or restaurants can be supplied, unless the grower plans to dry and package the herbs for later use. Other value-added products are also possible, including herbal tea mixtures, flavoured vinegars and oils, as well as packaged seeds.

Medicinal Herbs

The vast majority of medicinal herbs grown in Canada are produced organically. Aside from spearmint, which has had only regional production in Saskatchewan, and those herbs which are also regarded as spice or oil crops (such as borage and dill), only Echinacea has been the major herb crop to date. However, markets for medicinal herbs are very volatile and growers need to use caution before planting any medicinal herbs on a large scale. As with vegetables, medicinal herbs are mostly grown as row crops. Pest and disease problems are known for some and the incidence could increase if acreages become significant.

Other medicinal herbs that have been produced (mostly organically) in Saskatchewan include astragalus, burdock, calendula, catnip, chamomile, ellacampane, feverfew, licorice, marshmallow, milk thistle, motherwort, mullein, red clover, skullcap, St. John's wort, valerian and yarrow.

Even some weeds, such as stinging nettle and dandelion, may be marketable. Aster-yellows and Sclerotinia diseases have occasionally been serious problems in some of these crops.

Herb crops for cosmetic uses are also gaining prominence. Willow herb (fireweed) is one example. Much of it is gathered from the wild, rather than being cultivated.

Fruit Crops

The list of suitable fruit crops for organic production in Saskatchewan is rather limited:

- Sea buckthorn - This fruit does not have a fresh market, but has been marketed in processed form. It is often regarded as a nutraceutical product because of its high nutritional value. Harvesting and marketing are the main challenges with this crop. Suckering can also be a problem in established plantations. Development of thornless cultivars is deemed desirable for the prairie region.

- Sour cherries - The hardy cultivars developed by the University of Saskatchewan are showing much promise, and it should be possible to grow them organically. In time, pest control may be more difficult. Carmine Jewel was the first sour cherry released by the University of Saskatchewan in 1999. "Romance" series including "Juliet", "Crimson Passion", "Valentine", "Romeo" and "Cupid" were released in 2004. Various fruit flies, and possibly caterpillars, may need to be controlled on cherries.

- Haskap - An edible blue honeysuckle, *Lonicera coerulea*, is a relatively new fruit that shows promise. The Haskap has proved to be an easy-care shrub that has had use as a hardy ornamental shrub. Haskap is thought to be immune to the honeysuckle aphid, which often is very damaging to Tatarian types. A number of new Haskap varieties have been released from the University of Saskatchewan fruit program that show considerable promise for organic production.

- Apples - Have been grown organically in various parts of Canada and the USA. They are not without challenges. In recent years, apple maggot has become established in Saskatchewan and will have an impact on whether organic production is feasible here. Apple seed chalcid and fireblight can also cause problems. Apples have not been a traditional commercial crop on the prairies, but newer cultivars, with improved quality and hardiness, are making commercial ventures less risky. At present, organic production here should be considered only on a trial basis. Plums should also be considered in this category.

- Raspberries - Organic production may be somewhat difficult because of various pest problems, including raspberry crown borer, raspberry sawfly, raspberry fruit worm and spider mites.

- Strawberries - Pest control for numerous insects will likely be difficult for organic producers.

These include several plant bugs and strawberry root weevils. Diseases such as botrytis, causing fruit rot, can also be of concern.

- Saskatoon berries - This crop is a very poor candidate for organic production on the prairies. Pest control for Saskatoon berries is vital, and the lack of organic pesticides for control of pests such as Saskatoon bud moth and woolly elm aphids makes this crop a very risky venture except by conventional means. At best, production should be limited to very small scale operations.
- Other fruit - Black currants are gaining popularity and require some pest control measures. It may be possible to use red currants or gooseberries as lure crops so that the black currants will have a reduced incidence of pests such as currant fruit fly. Further research may be required to determine the effectiveness of this control method.

Other fruits such as high-bush cranberry and hardy grapes might be possible where a market can be established. Rhubarb, while technically a vegetable, is often considered with fruit crops. It usually has minimal pests and might be considered for organic production.

Organic Fruit Production

Organic fruit production essentially excludes the use of many inputs associated with modern farming, most notably synthetic pesticides and fertilizers. To the maximum extent possible, organic farming systems rely upon crop rotations, crop residues, animal manures, legumes, green manures, off-farm organic wastes, mechanical cultivation, mineral- bearing rock powders and biological pest control. These components maintain soil productivity and tilth, supply plant nutrients and help to control insects, weeds and other pests.

Cultural Practices

Site Selection

All factors regarding site suitability for conventional fruit plantings (air and water drainage, etc.) apply even more and so to organic operations. While conventional farmers may fall back on chemical fertilizers and pesticides to compensate for poor site decisions, organic farmers largely give up such luxuries. The presence of certain weeds and forage species are also of particular concern to the organic farmer. Bermudagrass, Johnsongrass and several other species can be quite problematic to farmers and are difficult to control through nonchemical means.

Growing fruit crops offers an advantage to farmers interested in sustainable agriculture. Because fruit plantings are perennial, the soil may not require additional tillage or cultivation beyond that needed at establishment, thereby minimizing soil erosion. Because the potential for erosion is low, hillsides and other sites unsuitable for tillage agriculture can safely and successfully produce fruit crops.

Crop Selection

Environmental constraints (climate, presence of certain pests and diseases, suitable soils, etc.) can greatly impact the suitability of a given site or even a bioregion for organic production of a given fruit crop. Generally speaking, some perennial fruits are easier to grow organically than others. The small fruits (berries) for example, seem easier to grow organically than the tree fruits in almost all locations. Lastly, successful organic fruit growing may depend largely on whether the venture is for home production or for commercial sales.

Site Preparation

Attention to the details of site preparation can go a long way towards reducing weed and disease problems and assuring a vital planting through soil improvement.

In general, fruit crops do not require highly fertile soils for good production. In fact, highly fertile soils, rich in nitrogen, can promote too much vegetative growth at the expense of fruiting. A nutritionally balanced soil, proper oil pH and plentiful amounts of organic matter are the fundamentals of an organic fertility management plan for fruits. Preplant soil improvement for organic fruit plantings is typically accomplished through some combination of cover cropping/green manuring and the use of imported materials, which may include manures, compost, rock powders and organic wastes.

Of particular importance at this stage is the application of required amounts of lime or sulphur. The need for lime or sulphur is dependent on the crop to be grown and the results of soil testing. Adjusting the oil pH with lime (to raise the pH) or sulphur (to lower pH) is much more easily done before planting. Most fruit plants perform best around pH 6.5, although they tolerate a pH range between 5.5 and 7.2. Blueberries are an exception. They require an acid oil-ideally pH 4.8 to 5.2 soil testing should also be used to guide applications of manures and other rock powders to avoid nutrient imbalances.

Cover crops and green manures not only contribute to soil fertility but also can become part of an active plan for pre-plant weed suppression as smother crops. The basic strategy begins with ploughing under sod or other existing vegetation, planting a cover crop to suppress weed growth, tilling under the cover crop and setting plants. Sometimes several rotations of cover crops are used before setting plants.

Selection of specific cover crops and their management varies with location, depending on such factors as seasonal rainfall, soil type, soil erosion potential, available equipment and seed cost. For example, crotalaria probably performed less effectively than sesbania at different centre because it is poorly adapted to low, wet soils. Some other warm season cover crops that might be considered include aggressive maize varieties and forage soybean varieties.

Crops such as berries (strawberries) are often grown on raised beds to encourage drainage and reduce root rot disease problems.

Soil Solarisation

Another technique for site preparation is soil solarization. The process involves placing transparent polyethylene plastic on moist soil during the hot summer months to increase soil temperatures to levels lethal to many pests.

Solarization suppresses weeds and eliminates many potential soil disease and nematode problems.

Soil solarization is somewhat expensive and is not frequently used to prepare sites for perennial fruits, with the possible exception of strawberries.

Orchard Floor Management and Mulching

The floor or inter row areas of perennial fruit plantings may be managed in a variety of ways, ranging from clean cultivation to various combinations of cover crops and organic mulches. In terms of controlling erosion at least, a system that results in full or close-to-full ground coverage is best. If the decision is made to plant a between row ground cover, it should consist of a species adapted to the region and to the fruit farmer's management plan. Where adapted, orchard grass and other cool season grasses are often recommended because they go dormant during the heat of the summer, thereby minimizing competition with the fruit crop for water. With proper fertility management these grasses may also provide plentiful material for mulch. Many warm season legumes are deep rooted and compete with the trees for water, they should not be allowed to grow under the tree canopy. However, legumes and the mulch made from leguminous ground covers can provide significant nitrogen to fruit trees or vines. Furthermore, even though they complete with the fruit plants for water, legumes (like the grasses) increase the water permeability of the soil. Increasing the organic matter content of the soil also increases the soil's moisture retentiveness.

Subterranean clover reseeds itself in early summer and dies back during hottest part of the growing season leaving a relatively thick, weed suppressive duff or mulch. This system has shown applicability in apple and peach orchards and for a variety of orchard crops. Subterranean clover is not adapted to climates where winter temperatures regularly drop below 0 °F.

Many researchers have also investigated the relationship between orchard floor management and pest control. In some place apple orchards peach orchards, a diverse mix of cover crop spices provided habitat and food for an array of beneficial organisms, resulting in a decrease of orchard pests.

Mustards, buckwheat, dwarf sorghum and various members of the Umbelliferae and Compositae families support substantial numbers of beneficial insects without attracting as many pests.

After a planting is established, mulching with organic materials such as straw, leaves or sawdust can provide significant weed suppression. If applied thickly enough or supplemented with sheets of paper or cardboard as the bottom layer, complete suppression might be achieved. Application of mulch varies somewhat by crop. In strawberries for instance, mulch might cover entire aisles between rows or only be placed next to the bed to inhibit encroachment by creeping weeds. In vineyards and orchards, mulch might be placed only under individual trees or vines or along the entire row, ideally extending to the drip line.

Generally, mulch should be kept well away from the trunk to reduce damage from voles. This is especially important in winter. Keeping the mulch 8-12 inches away from the trunk also reduces the likelihood of crown rot and other diseases in susceptible species most notably applies on certain rootstocks. In mulberries, sawdust mulch is commonly spread along the entire row with extra sawdust mounded around the canes, often to a depth of 8 or more inches. In addition to controlling weeds, mulching with organic materials improves the soil by enhancing soil aggregation and water availability.

Varietal Selection

Because the plants are perennial and represent a considerable investment in both time and money, it is important to start the fruit planting with the optimum varieties for location and markets.

Good information on varietal selection is available from Cooperative Extension, nurseries and local commercial farmers. It is also important to obtain clean planting stock. Buying from reputable nurseries that provide stock certified by certifying inspectors to be free of diseases and insect pests is recommended.

Genetic resistance refers to inheritable traits in the plants that inhibit disease and pest damage. Choosing genetically resistant cultivars is a very important control measure for organic farmers especially with regard to disease management. In some cases, such as bacterial spot in peaches, cultivar resistance is the best or only control.

Aphid resistant berries, Phylloxera resistant grape rootstocks, wooly aphid resistant apple rootstocks, mite resistant strawberry cultivars, and nematode resistant peach rootstocks are available. As important as this resistance is there is no cultivar of any fruit species with multiple insect pest resistance; therefore, means other than resistance will most likely have to be employed to protect fruit plant from a complex of several pest species.

Organic Fertilization Practices

Harvested fruit, being largely water, removes relatively few nutrients from the soil, compared to other crops. Therefore, a significant amount of the fertility needs of fruit crops can be met through cover crop management and organic mulches in systems which use them and by the application of lime and other slow release rock powders at the preplant stage. Still, supplementary fertilization is often required for optimum growth and production.

Some information useful to planning a supplementary fertilization program for perennial fruits include:

- Organic farmers generally employ relatively non-soluble fertilizer materials, such as compost, manures, plant derived byproducts like cottonseed meal or animal byproducts like feather or blood meals. To insure adequate decomposition and timely nutrient release of these slowly available materials, early spring application is encouraged. Early application also reduces the tendency towards late season growth, which may result in winter damage.
- Surface application of organic fertilizer materials without incorporation is sometimes wasteful of the nitrogen contained in those materials. This is especially true of manures. However, incorporation by tillage could damage the roots of the fruit plants and increase erosion.
- To bypass some of the "problems" associated with slowly available organic materials, some organic fruit farmers choose soluble organic fertilizers such as fish emulsion, soluble fish

powder or water soluble blood meal in some cases applying these as a foliar spray. Since these are relatively expensive, the prudent farmer seldom relies exclusively on such materials.

- Most organic fertilization programs focus on supplementing nitrogen as the key element since it is needed in the greatest amount for the crop. One way to determine a proper application rate for an organic fertilizer is to obtain a conventional chemical recommendation for the planting and calculate the approximate amount of an organic fertilizer required to meet it. For instance, if the recommendation for a berry crop is 160 kg of actual nitrogen per acre, a farmer choosing to use cottonseed meal (approximately 7% N) would need to apply about 2000 kg of that material to each acre. Such calculations are simplistic however and may lead to spending more on fertilizer inputs than in necessary. One reason is because biologically healthy soils fix and release greater amounts of nitrogen naturally than those which are not.
- Further more, when making fertilizer calculations based on nitrogen, the farmer needs to credit the estimated contributions made by legume cover crops and or leguminous mulches, where these are used. A cover crop of subterranean clover, properly fertilized and inoculated, can fix from 200 to 1000 kg of nitrogen per acre annually in a living mulch system.
- Basing application rates solely on nitrogen content can cause problems when the fertilizers themselves are imbalanced. Repeated use of poultry manure for example which is very high in phosphate, can lead to pollution problems and a zinc deficiency in the crop. These problems can be averted by regular soil testing and adjusting fertilizer selection and rates accordingly.
- The most reliable means for determining whether fertilization is adequate is to combine field observations with soil and tissue testing. Poor yields, unusual colouration of leaves, and poor plant growth are all clues to a possible nutritional imbalance or deficiency. For example, less than 10 inches annual elongation of the branches of most fruit trees probably indicates a nitrogen deficiency. A corky bark on certain apple varieties may indicate an over availability of manganese in the soil.
- A foliar analysis measures the nutrient content of the leaves and can identify a deficiency or excess well in advance of visible symptoms. It is more helpful than a soil test because the foliar analysis is a measure of what the plant is actually taking up, while soil analysis only measures what is in the soil, which may or may not be available to the plant. Annual foliar analyses generally provide the best guide for adjusting supplementary nitrogen fertilization.

Organic Weed Management

Research indicates that without some form of weed control in the fruit planting, crop yields and plant vigour will be greatly reduced. In organic farming, weed control is only one goal of a weed management system for perennial fruit crops. A good organic weed management plan should present a minimum erosion risk, provide a "platform" for the movement of farm equipment, not impact adversely on pest management or soil fertility, while minimizing weed competition for water and nutrients. This philosophy has already been demonstrated in discussions regarding three effective weed control tools: cover crops, mulches and soil solarization.

Organic Insect and Mite Pest Management

A major distinction between pest management in perennial fruit crops and in annual crops is that crop rotation is not an option (strawberries and to a lesser extent, brambles, are possible exceptions). The long term nature of fruit growing allows for the possible build up of a pest population over time. Conversely, it is also possible for such stable agricultural environments to sustain populations of beneficial organisms.

Plant Health and Vigour

Though it is sometimes overstated, maintaining the plant in general health and good vigour is important in pest management. For fruit plants, this adage is more applicable to indirect pests (those pests that feed on foliage, stems etc.) than to direct pests (pests that feed on the fruit). For instance, an apparently healthy plum tree may set a good crop of fruit, yet lose it all to the plum. On the other hand, the same tree might suffer significant defoliation by caterpillars early in the season; yet, if it is in good vigour, it can compensate and bounce back quickly still producing a marketable crop that year. There are some cases where general plant health and freedom from stress does impart a form of "resistance" not technically genetic resistance to certain pests. Two examples are apple trees in good vigour actually smothering with sap or casting out invading flathead apple tree borers and plants not suffering drought stress being much less attractive to grasshoppers.

Biological Control

Biological control refers to the use of living organisms to control the population of a pest. Examples of beneficial arthropods that have been used to control pests in fruit crops include the predatory mites Phytoseiulus persimilis and Metaseiulus occidentalis, which attack spider mites; lady bird beetles and green lacewings which feed on aphids and Trichogramma wasps, which parasitize the eggs of several pests including codling moth.

Many beneficial insects can be purchased from commercial insectaries and released in fruit plantings. More economical, however is the management of cover crops and adjacent vegetation as insect refugia to attract and sustain native populations of beneficials. As a rule, it appears that beneficial arthropods are not a complete control measure for direct fruit pests at least for commercial farmers who have a low damage threshold for fresh fruit. Usually, additional control measures are required.

Organic and Biorational Pesticides

Pesticides approved for certified organic production are usually derived from natural sources, break down rapidly in the field and appear to have minimal impact on the environment. Examples include botanical extracts from plants, insect growth regulators, synthetic pheromone treatments that cause mating disruption, soaps, oils, minerals such as sulphur dust and biological pesticides. The term "biorational pesticide" is used to describe pesticides organically acceptable or synthetic, which have minimal impact on beneficial insects and the environment.

Biorational pesticides are considered those providing the least toxic control that can be applied against a pest. Biological pesticides (biopesticides) are both organically acceptable and biorational. Biopesticides differ from biological control in part because they are formulated, labeled and applied like standard pesticides but also because the organisms involved do not reproduce significantly in the field. Biopesticides are highly specific and do not harm humans or beneficial insects. Several biopesticides can be used in fruit pest management. To be effective, they must be used at a specific time in the pest's development cycle.

The bacterium Bacillus thuringiensis (Bt) is an example of a commonly used biological insecticide. Bt is not as effective against lepidopterous pests that spend their larval stage feeding inside stems, crowns, trunks, or fruit, etc. (e.g., peach tree borer, codling moth, grape root borer, etc.). Other microbial insecticides include Bacillus popilliae for Japanese beetle grubs, a granulosis virus for codling moth, and insect parasitic nematodes for grubs and wireworms.

Botanical insecticides are formulated by extracting toxic compounds from plants that have pesticidal properties. They are naturally occurring, short-lived in the environment and do not leave harmful residues. However, many botanicals are broad spectrum poisons, affecting pests and beneficial organisms alike and are not always the biorational choice. Organic farmers, who are prohibited from using synthetic pesticides, frequently use botanical extracts. Some commonly used plant-derived insecticides are pyrethrum, rotenone, ryania and neem.

Specially formulated soaps that are high in fatty acids are effective against several soft bodied insects including aphids, whiteflies, leafhoppers and spider mites. Insecticidal soap penetrates the insect's body and disrupts the normal function of cells and their membranes, causing the contents to leak out.

Applying a thin layer of dormant oil to certain woody plants such as fruit trees, grapevines and bushes suppresses pests like leaf rollers, aphids, mites and scales by suffocating over wintering adults and eggs. Dormant oils should be applied prior to bud break and should never be mixed with sulphur because foliage damage may occur. So-called summer oils have a very low viscosity and can be used during the growing season to control aphids and mites without plant damage if label cautions are observed. Insect pheromones are chemicals produced by insects to help them communicate such things as mate availability and sexual receptivity. They are usually specific to a given insect species or genus.

Scientists have learned how to synthesize many of these pheromones and they are widely used for monitoring the emergence or simple presence of crop pests. This information is commonly used to time pesticide applications. New technology also allows pheromones to be used for mating disruption of certain pests. Mating disruption pheromones are available for the oriental fruit moth, codling moth and peach tree borer and grape berry moth.

Organic Disease Management

General Disorders and Cultural Controls

There are however, some types of disease problems, which are common to almost all temperate zone perennial fruit crops. For instance, because of the relatively soft nature and high sugar content of most mature or nearly mature fruits, fruit rots are common afflictions. The organic farmer can help to minimize the chances of fruit rots by allowing good air circulation and sunlight penetration into the interior plant canopy. Sunlight and circulating air help to dry leaf and fruit surfaces, thereby limiting fungal and bacterial infections. In tree crops, this would mean proper pruning and training techniques. In brambles and berries, reducing plant density helps. In grapes, discouraging rank vine growth and removing leaves that shade fruit clusters is beneficial. For all fruit crops, a site that allows for good air circulation should be chosen.

Another problem common to many fruit crops is root rots and intolerance of poorly drained soils. Most pear rootstocks and some apple rootstocks are relatively tolerant of heavy of poorly drained soils, but even these crops will succumb to persistently waterlogged conditions. Prunus species (peaches, plums, cherries, etc) are very intolerant of poorly drained soils and are generally susceptible to root rotting organisms common in such soils. Other cultural aids in minimizing disease can include such things as maintaining plants in good health and vigour, removal through prunings from the planting site, rouging out of diseased plants, removal of alternate hosts or inoculum sources for the disease organisms.

Organic Fungicides and Bactericides

Copper and sulphur compounds are the principal fungicides and bactericides used by organic farmers, but they have drawbacks. These materials can cause damage to plants if applied incorrectly. Sulphur is also lethal to some beneficial insects, spiders and mites and can set the stage for further pest problems. Long term use of copper fungicides can also lead to toxic levels of copper in the soil. Furthermore, these fungicides are typically inferior to synthetic alternatives, and have to be used on a protective schedule requiring frequent applications. Research on biofungicides is encouraging. Several formulations of the fungus Trichoderma harzianium are now come to market as a control for grey mold (Botrytis). Other biofungicides now available include a control for powdery mildew in grapes and a protectant against tree wound pathogens.

Organic Food

Organic food is the fresh or processed food produced by organic farming methods. Organic food is grown without the use of synthetic chemicals, such as human-made pesticides and fertilizers, and does not contain genetically modified organisms (GMOs). Organic foods include fresh produce, meats, and dairy products as well as processed foods such as crackers, drinks, and frozen meals. The market for organic food has grown significantly since the late 20th century, becoming a multibillion dollar industry with distinct production, processing, distribution, and retail systems.

Policy

Although organic food production began as an alternative farming method outside the mainstream, it eventually became divided between two distinct paths: (1) small-scale farms that may not be formally certified organic and thus depend on informed consumers who seek out local, fresh, organically grown foods; and (2) large-scale certified organic food (fresh and processed) that is typically transported large distances and is distributed through typical grocery store chains. If consumers know their local farmer and trust the farmer's production methods, they may not demand a certification label. On the other hand, organic food produced far away and shipped is more likely to require a certification label to promote consumer trust and to prevent fraud, which exemplifies how national certification regulations are most beneficial.

A regulatory framework is most important when consumers and farmers are geographically separated, and such a framework is likely to cater to larger-scale producers who participate in a more industrial system. This regulatory approach does not necessarily match consumers' assumptions about organic food production, which typically include images of small family farms and the humane treatment of animals. In general, regulations surrounding organic food do not address more complex social concerns about family farms, farmworker wages, or farm size, and organic policy in some places does little to address animal welfare.

Organic food policies were created largely to provide a certification system with specific rules regarding production methods, and only products that follow the guidelines are allowed to use the certified organic labels. In the United States, the Organic Foods Production Act of 1990 began the process of establishing enforceable rules to mandate how agricultural products are grown, sold, and labeled. The regulations concerning organic food and organic products are based on a National List of Allowed and Prohibited Substances, which is a critical aspect of certified organic farming methods. The United States Department of Agriculture (USDA) regulates organic production through its National Organic Program (NOP), which serves to facilitate national and international marketing and sales of organically produced food and to assure consumers that USDA certified organic products meet uniform standards. To this end, NOP established three specific labels for consumers on organic food products: "100% organic," "organic," or "made with organic," which signify that a product's ingredients are 100 percent, at least 95 percent, or 70 percent organic, respectively. Noncertified products cannot use the USDA organic seal, and violators face significant fines and penalties.

Organic regulations vary by country, some of the most comprehensive rules being seen in Europe. Objectives of organic farming in the European Union (EU) include respecting nature's biological systems and establishing a sustainable management system; using water, soil, and air responsibly; and adhering to animal welfare standards that meet species-specific behavioral needs. In addition, principles of organic production in the EU are based on designing and managing farms to promote ecological systems and on using natural resources within the farming system. These policy goals go far beyond a defined listing of prohibited materials in organic production.

Environment

The overall impacts of organic agriculture are beneficial to the environment. Certified organic production methods prohibit the use of synthetic fertilizers and pesticides, thus reducing chemical runoff and the pollution of soils and watersheds. Smaller-scale organic farming often is associated with significant environmental benefits, owing to the use of on-farm inputs, such as fertilizers derived from compost created on-site. By comparison, large-scale organic farms often require inputs generated off-site and may not employ integrated farming methods. These operations may buy specific allowable inputs, such as fish emulsion or blood meal to use as fertilizer rather than working within the farm to increase soil fertility. While this decrease in synthetic chemical use benefits the environment compared with industrial agriculture, these methods may not promote long-term sustainability, since off-farm inputs usually require greater fossil fuel use than on-farm inputs.

Society

Social concerns related to organic food include higher costs to consumers and geographic variations in demand. Organic food usually is more expensive for consumers than conventionally produced food because of its more labour-intensive methods, the costs of certification, and the decreased reliance on chemicals to prop up crop yields. This often translates into unequal access to organic food. Research indicates that greater wealth and education levels are correlated with organic food purchases. Further, there are trends in some lower-income countries to produce certified organic crops solely for export to wealthier countries. This sometimes generates a situation in which the farmers themselves cannot afford to buy the organic foods they are producing. While this strategy may bring economic gain in the short term, it is a concern when farmers are forced out of producing food crops that feed their local communities, thus increasing food insecurity.

Certified organic agriculture has also become a big business in many places, with larger farming operations playing a key role in national and global certified organic food markets. Given economies of scale, big food-processing companies often buy from a single farming operation that produces organic crops on thousands of acres, rather than from many smaller farms that each grow on smaller acreages, a practice that effectively limits the participation of smaller farmers in these markets. There also is disparity among farmers, since the organic certification process can be prohibitively expensive to some smaller-scale farmers. Although certification subsidies exist in some places, such farmers often opt to sell directly to consumers at farmers' markets, for example, and may decide to forgo organic certification altogether.

Overall, organic food has grown in popularity, as consumers have increasingly sought and purchased foods that they perceive as being healthier and grown in ways that benefit the environment. Indeed, consumers typically buy organic food in order to reduce their exposure to pesticide residues and GMOs. Further, some research shows that organically produced crops have higher nutritional content than comparable nonorganic crops, and some people find organic foods to be tastier. The question remains, however, whether organic food shipped in from across the globe is truly a sustainable method of food production. Certainly organically produced food from a local farmer who employs an integrated whole-farm approach is fairly environmentally sustainable, though the economic sustainability of such an endeavour can be challenging. Although humans must decrease their reliance on fossil fuels to combat climate change, many organic policies do little to address the issue of sustainability, focusing instead on the strict list of prohibited substances, rather than a comprehensive long-term view of farming and food.

References

- Organic-farming-horticulture: iitk.ac.in, Retrieved 29 March, 2019
- Start-an-organic-garden: conserve-energy-future.com, Retrieved 16 January, 2019
- Organic-production-of-horticultural-crops, organic-crops, crops-and-irrigation, agribusiness-farmers-and-ranchers, agriculture-natural-resources-and-industry, business: saskatchewan.ca, Retrieved 30 August, 2019
- Orgfarm-fruits, or-farm: tnau.ac.in, Retrieved 15 July, 2019
- Organic-food, topic: britannica.com, Retrieved 30 April, 2019

Gardening 5

- **Landscape Gardening**
- **Container Gardening**
- **Sustainable Gardening**
- **Bonsai Cultivation and Care**

The practice of cultivating and growing plants as part of horticulture is known as gardening. Some of the major types of gardening are landscape gardening, container gardening, bonsai cultivation and sustainable gardening. This chapter has been carefully written to provide an easy understanding of these types of gardening.

Gardening is the laying out and care of a plot of ground devoted partially or wholly to the growing of plants such as flowers, herbs, or vegetables.

Gardening can be considered both as an art, concerned with arranging plants harmoniously in their surroundings, and as a science, encompassing the principles and techniques of plant cultivation. Because plants are often grown in conditions markedly different from those of their natural environment, it is necessary to apply to their cultivation techniques derived from plant physiology, chemistry, and botany, modified by the experience of the planter. The basic principles involved in growing plants are the same in all parts of the world, but the practice naturally needs much adaptation to local conditions.

The Nature of Gardening

Gardening in its ornamental sense needs a certain level of civilization before it can flourish. Wherever that level has been attained, in all parts of the world and at all periods, people have made efforts to shape their environment into an attractive display. The instinct and even enthusiasm for gardening thus appear to arise from some primitive response to nature, engendering a wish to produce growth and harmony in a creative partnership with it.

It is possible to be merely an admiring spectator of gardens. However, most people who cultivate a domestic plot also derive satisfaction from involvement in the processes of tending plants. They find that the necessary attention to the seasonal changes, and to the myriad small "events" in any shrubbery or herbaceous border, improves their understanding and appreciation of gardens in general.

A phenomenal upsurge of interest in gardening began in Western countries after World War II. A lawnwith flower beds and perhaps a vegetable patch has become a sought-after advantage to home ownership. The increased interest produced an unprecedented expansion of business among horticultural suppliers, nurseries, garden centres, and seedsmen. Books, journals, and newspaper columns on garden practice have found an eager readership, while television and radio programs on the subject have achieved a dedicated following.

Several reasons for this expansion suggest themselves. Increased leisure in the industrial nations gives more people the opportunity to enjoy this relaxing pursuit. The increased public appetite for self-sufficiency in basic skills also encourages people to take up the spade. In the kitchen, the homegrown potato or ear of sweet corn rewards the gardener with a sense of achievement, as well as with flavour superior to that of store-bought produce. An increased awareness of threats to the natural environment and the drabness of many inner cities stir some people to cultivate the greenery and colour around their own doorsteps. The bustle of 20th-century life leads more individuals to rediscover the age-old tranquillity of gardens.

The Varied Appeal of Gardening

The attractions of gardening are many and various and, to a degree perhaps unique among the arts and crafts, may be experienced by any age group and at all levels of ambition. At its most elemental, but not least valuable, the gardening experience begins with the child's wonder that a packet of seeds will produce a charming festival of colour. At the adult level, it can be as simple as helping to raise a good and edible carrot, and it can give rise to almost parental pride. At higher levels of appreciation, it involves an understanding of the complexity of the gardening process, equivalent to a chess game with nature, because the variables are so many.

The gardening experience may involve visiting some of the world's great gardens at different seasons to see the relation of individual groups of plants, trees, and shrubs to the whole design; to study the positioning of plants in terms of their colour, texture, and weight of leaf or blossom; and to appreciate the use of special features such as ponds or watercourses, pavilions, or rockeries. Garden visiting on an international scale provides an opportunity to understand the broad cultural influences, as well as the variations in climate and soil, that have resulted in so many different approaches to garden making.

The appeal of gardening is thus multifaceted and wide in range. The garden is often the only place where someone without special training can exercise creative impulses as designer, artist, technician, and scientific observer. In addition, many find it a relaxing and therapeutic pursuit. It is not surprising that the garden, accorded respect as a part of nature and a place of contemplation, holds a special place in the spiritual life of many.

Practical and spiritual aspects of gardening are shown in an impressive body of literature. In Western countries manuals of instruction date to classical Greece and Rome. Images of plants and gardens are profuse in the works of the major poets, from Virgil to Shakespeare, and on to some of the moderns.

Another of gardening's attractions is that up to a certain level it is a simple craft to learn. The beginner can produce pleasing results without the exacting studies and practice required by, for example, painting or music. Gardens are also forgiving to the inexperienced to a certain degree. Nature's exuberance will cover up minor errors or short periods of neglect, so gardening is an art practiced in a relatively nonjudgmental atmosphere. While tolerant in many respects, nature does, however, present firm reminders that all gardening takes place within a framework of natural law; and one important aspect of the study of the craft is to learn which of these primal rules are imperatives and which may be stretched.

Control and Cooperation

Large areas of gardening development and mastery have concentrated on persuading plants to achieve what they would not have done if left in the wild and therefore "natural" state. Gardens at all times have been created through a good deal of control and what might be called interference. The gardener attends to a number of basic processes: combating weeds and pests; using space to allay the competition between plants; attending to feeding, watering, and pruning; and conditioning the soil. Above this fundamental level, the gardener assesses and accommodates the unique complex of temperature, wind, rainfall, sunlight, and shade found within his own garden boundaries. A major part of the fascination of gardening is that in problems and potential no one garden is quite like another; and it is in finding the most imaginative solutions to challenges that the gardener demonstrates artistry and finds the subtler levels of satisfaction.

Different aesthetics require different balances between controlling nature and cooperating with its requirements. The degree of control depends on the gardener's objective, the theme and identity he is aiming to create. For example, the English wild woodland style of gardening in the mid-19th century dispensed with controls after planting, and any interference, such as pruning, would have been misplaced. At the other extreme is the Japanese dry-landscape garden, beautifully composed of rock and raked pebbles. The artistic control in this type of garden is so firm and refined that the intrusion of a single "natural" weed would spoil the effect.

Choice of Plants

The need for cooperation with nature is probably most felt by the amateur gardener in choosing the plants he wants to grow. The range of plants available to the modern gardener is remarkably rich, and new varieties are constantly being offered by nurseries. Most of the shrubs and flowers used in the Western world are descendants of plants imported from other countries. Because they are nonnative, they present the gardener with some of his most interesting problems but also with the possibility of an enhanced display. Plants that originated in subtropical regions, for example, are naturally more sensitive to frost. Some, like rhododendrons or azaleas, originated in an acid soil, mainly composed of leaf mold. Consequently, they will not thrive in a chalky or an alkaline soil. Plant breeding continues to improve the adaptability of such exotic plants, but the more closely the new habitat resembles the original, the better the plant will flourish. Manuals offer solutions to most such problems, and the true gardener will always enjoy finding his own. In such experiments, he may best experience his work as part of the historical tradition of gardening.

Types Of Gardens

The domestic garden can assume almost any identity the owner wishes within the limits of climate, materials, and means. The size of the plot is one of the main factors, deciding not only the scope but also the kind of display and usage. Limits on space near urban centres, as well as the wish to spend less time on upkeep, have tended to make modern gardens ever smaller. Paradoxically, this happens at a time when the variety of plants and hybrids has never been wider. The wise small gardener avoids the temptations of this banquet. Some of the most attractive miniature schemes, such as those seen in Japan or in some Western patio gardens, are effectively based on an austere simplicity of design and content, with a handful of plants given room to find their proper identities.

In the medium- to large-sized garden, the tradition generally continues of dividing the area to serve various purposes: a main ornamental section to enhance the residence and provide vistas; walkways and seating areas for recreation; a vegetable plot; a children's play area; and features to catch the eye here and there. Because most gardens are mixed, the resulting style is a matter of emphasis rather than exclusive concentration on one aspect. It may be useful to review briefly the main garden types.

Flower Gardens

Though flower gardens in different countries may vary in the types of plants that are grown, the basic planning and principles are nearly the same, whether the gardens are formal or informal. Trees and shrubs are the mainstay of a well-designed flower garden. These permanent features are usually planned first, and the spaces for herbaceous plants, annuals, and bulbs are arranged around them. The range of flowering trees and shrubs is enormous. It is important, however, that such plants be appropriate to the areas they will occupy when mature. Thus it is of little use to plant a forest tree that will grow 100 feet (30 metres) high and 50 feet across in a small suburban front garden 30 feet square, but a narrow flowering cherry or redbud tree would be quite suitable.

Blending and contrast of colour as well as of forms are important aspects to consider in planning a garden. The older type of herbaceous border was designed to give a maximum display of colour in summer, but many gardeners now prefer to have flowers during the early spring as well, at the expense of some bare patches later. This is often done by planting early-flowering bulbs in groups toward the front. Mixed borders of flowering shrubs combined with herbaceous plants are also popular and do not require quite so much maintenance as the completely herbaceous border.

Groups of half-hardy annuals, which can withstand low night temperatures, may be planted at the end of spring to fill gaps left by the spring-flowering bulbs. The perpetual-flowering roses and some of the larger shrub roses look good toward the back of such a border, but the hybrid tea roses and the floribunda and polyantha roses are usually grown in separate rose beds or in a rose garden by themselves.

Woodland Gardens

The informal woodland garden is the natural descendant of the shrubby "wilderness" of earlier times. The essence of the woodland garden is informality and naturalness. Paths curve rather than run straight and are of mulch or grass rather than pavement. Trees are thinned to allow enough light, particularly in the glades, but irregular groups may be left, and any mature tree of character can be a focal point. Plants are chosen largely from those that are woodlanders in their native countries: rhododendron, magnolia, pieris, and maple among the trees and shrubs; lily, daffodil, and snowdrop among the bulbs; primrose, hellebore, St.-John's-wort, epimedium, and many others among the herbs.

Rock Gardens

Rock gardens are designed to look as if they are a natural part of a rocky hillside or slope. If rocks are added, they are generally laid on their larger edges, as in natural strata. A few large boulders usually look better than a number of small rocks. In a well-designed rock garden, rocks are arranged so that there are various exposures for sun-tolerant plants such as rockroses and for shade-tolerant plants such as primulas, which often do better in a cool, north-facing aspect. Many smaller perennial plants are available for filling spaces in vertical cracks among the rock faces.

The main rocks from which rock gardens are constructed are sandstone and limestone. Sandstone, less irregular and pitted generally, looks more restful and natural, but certain plants, notably most of the dianthuses, do best in limestone. Granite is generally regarded as too hard and unsuitable for the rock garden because it weathers very slowly.

Water Gardens

The water garden represents one of the oldest forms of gardening. Egyptian records and pictures of cultivated water lilies date as far back as 2000 BCE. The Japanese have also made water gardens to their own particular and beautiful patterns for many centuries. Many have an ornamental lantern of stone in the centre or perhaps a flat trellis roof of wisteria extending over the water. In Europe and North America, water gardens range from formal pools with rectangular or circular outline, sometimes with fountains in the centre and often without plants or with just one or two water lilies (Nymphaea), to informal pools of irregular outline planted with water lilies and other water plants and surrounded by boggy or damp soil where moisture-tolerant plants can be grown. The pool must contain suitable oxygenating plants to keep the water clear and support any introduced fish. Most water plants, including even the large water lilies, do well in still water two to five feet deep. Temperate water lilies flower all day, but many of the tropical and subtropical ones open their flowers only in the evening.

In temperate countries water gardens also can be made under glass, and the pools can be kept heated. In such cases, more tropical plants, such as the great Victoria amazonica (V. regia) or the lotus (Nelumbo nucifera), can be grown together with papyrus reeds at the edge. The range of moisture-loving plants for damp places at the edge of the pool is great and includes many beautiful plants such as the candelabra primulas, calthas, irises, and osmunda ferns.

Herb and Vegetable Gardens

Most of the medieval gardens and the first botanical gardens were largely herb gardens containing plants used for medicinal purposes or herbs such as thyme, parsley, rosemary, fennel, marjoram, and dill for savouring foods. The term herb garden is usually used now to denote a garden of herbs used for cooking, and the medicinal aspect is rarely considered. Herb gardens need a sunny position, because the majority of the plants grown are native to warm, dry regions.

The vegetable garden also requires an open and sunny location. Good cultivation and preparation of the ground are important for successful vegetable growing, and it is also desirable to practice a rotation of crops as in farming. The usual period of rotation for vegetables is three years; this also helps to prevent the carryover from season to season of certain pests and diseases.

The old French potager, the prized vegetable garden, was grown to be decorative as well as useful; the short rows with little hedges around and the high standard of cultivation represent a model of the art of vegetable growing. The elaborate parterre vegetable garden at the Château de Villandry is perhaps the finest example in Europe of a decorative vegetable garden.

Specialty Gardens

Roof Gardens

The modern tendency in architecture for flat roofs has made possible the development of attractive roof gardens in urban areas above private houses and commercial buildings. These gardens follow the same principles as others except that the depth of soil is less, to keep the weight on the rooftop low, and therefore the size of plants is limited. The plants are generally set in tubs or other containers, but elaborate roof gardens have been made with small pools and beds. Beds of flowering plants are suitable, among which may be stood tubs of specimen plants to produce a desired effect.

Scented Gardens

Scent is one of the qualities that many people appreciate highly in gardens. Scented gardens, in which scent from leaves or flowers is the main criterion for inclusion of a plant, have been established, especially for the benefit of blind people. Some plants release a strong scent in full sunlight, and many must be bruised or rubbed to yield their fragrance. These are usually grown in raised beds within easy reach of visitors.

Contents of Gardens

Permanent Elements

The more or less permanent plants available for any garden plan are various grasses for lawns, other ground-cover plants, shrubs, climbers, and trees. More transitory and therefore in need of continued attention are the herbaceous plants, such as the short-lived annuals and biennials, and the perennialsand bulbous plants, which resume growth each year.

Lawns and Ground Covers

Areas of lawn, or turf, provide the green expanse that links all other garden plantings together. The main grasses used in cool areas for fine-textured lawns are fescues (Festuca species), bluegrasses (Poaspecies), and bent grasses (Agrostis species), often in mixtures. A rougher lawn mixture may contain ryegrass (Lolium species). In drier and subtropical regions, Bermuda grass (Cynodon dactylon) is frequently used, but it does not make nearly as fine a lawn as those seen in temperate regions of higher rainfall.

Ground covers are perennial plants used as grass substitutes in regions where grasses do poorly, or they are sometimes combined with grassy areas to produce a desired design. The deep greens, bronzes, and other colours that ground-cover plants can provide offer pleasing contrasts to the green of a turf. Ground covers, however, are not so durable as lawns and do not sustain themselves as well under foot traffic and other activities. Among the better known plants used as ground covers are Japanese spurge (Pachysandra terminalis), common periwinkle (Vinca minor), lily of the valley (Convallaria majalis), ajuga, or bugleweed (Ajuga reptans), many stonecrops (Sedum species), dichondra (Dichondra repens), and many ivies (Hedera species).

Shrubs and Vines

Smaller woody plants, such as shrubs and bushes, have several stems arising from the base. These plants attain heights up to about 20 feet (6 metres). They often form the largest part of modern gardens, because their cultivation requires less labour than that of herbaceous plants, and some flowering shrubs have extended blooming periods. Among the popular garden shrubs are lilac (Syringa vulgaris), privet (Ligustrum species), spirea (Spiraea species), honeysuckle (Lonicera species), forsythia (Forsythia species), mock orange (Philadelphus species), and hydrangea (Hydrangeaspecies).

Bushlike azaleas and rhododendrons (both of which are species of *Rhododendron*) provide colourful blossoms in spots where there is semishade.

Climbers are often useful in softening the sharp lines of buildings, fences, and other structures. They can provide shade as an awning or cover on an arbour or garden house. Some species are also useful as ground covers on steep slopes and terraces. Among the many woody perennial climbers for the garden are the ivies, trumpet creeper (*Bignonia*, or *Campsis, radicans*), clematis (*Clematis* species), wisteria (*Wisteria sinensis*), climbing roses, annual herbaceous vines such as morning glory (*Ipomoea*species), and ornamental gourds, the last of which can provide rapid but temporary coverage of unsightly objects.

Trees

Trees are the most permanent features of a garden plan. The range of tree sizes, shapes, and colours is vast enough to suit almost any gardening scheme, from shrubby dwarf trees to giant shade trees, from slow to rapid growers, from all tones of green to bronzes, reds, yellows, and purples. A balance between evergreen trees, such as pines and spruces, and deciduous trees, such as oaks, maples, and beeches, can provide protection and visual interest throughout the year.

Transitory Elements

Herbaceous Plants

Herbaceous plants, which die down annually and have no woody stem aboveground, are readily divided into three categories, as mentioned earlier: (1) Annuals, plants that complete their life cycle in one year, are usually grown from seed sown in the spring either in the place they are to flower or in separate containers, from which they are subsequently moved into their final position. Annuals flower in summer and die down in winter after setting seed. Many brilliantly coloured ornamental plants as well as many weeds belong in this category. Examples of annuals are petunia and lobelia. (2) Biennialsare plants sown from seed one year, generally during the summer. They flower the second season and then die. Examples are wallflower and sweet william. (3) Herbaceous perennials are those that die down to the ground each year but whose roots remain alive and send up new top growth each year. They are an important group in horticulture, whether grown as individual plants or in the assembly of the herbaceous border. Because they flower each year, they help to create the structure of the garden's appearance, so their placement must be considered carefully. Examples are delphinium and lupine.

Bulbous Plants

For horticultural purposes, bulbous plants are defined to include those plants that have true bulbs (such as the daffodil), those with corms (such as the crocus), and a few that have tubers or rhizomes (such as the dahlia or iris). A bulb is defined as a modified shoot with a disklike basal plate and above it a number of fleshy scales that take the place of leaves and contain foods such as starch, sugar, and some proteins. Each year a new stem arises from the centre. A corm consists of the swollen base of a stem, generally rounded or flattened at the top and covered with a membranous tunic in which reserve food materials are stored. A tuber or rhizome is not the base of the stem but rather a swollen part of an underground stem; it is often knobbly. All such plants have evolved in places where they can survive in a semidormant state over long unfavourable seasons, either cold mountain winters or long droughty summers.

The Principles of Gardening

Soil: Its Nature and Needs

Soil is the basic element in the cultivation of all plants, although soilless growth in water, with or without gravel or sand, enriched with suitable chemicals (hydroponics) can be very successful.

Soil consists of particles, mainly mineral, derived from the breakdown of rocks and other substances together with organic matter. In the pore spaces between the particles, both water (containing dissolved salts) and air circulate. The air contains more carbon dioxide and less oxygen than does the atmosphere. Minute living organisms are also present in soil in immense quantities and are what make it "alive." Plants must penetrate this pore space to reach much of their nourishment.

The soil must be managed for fertility (the ability to supply plant nutrients) and physical condition. Nutrients must be supplied and released in forms available to the plant. Sixteen elements are necessary for plant growth. Three of these, carbon, oxygen, and hydrogen, are provided through water and air; the other 13 are provided through the soil. The elements required in relatively large amounts are called major elements: nitrogen, phosphorus, potassium, calcium, magnesium, and sulfur. The minerals required in small quantities are called trace elements: iron, boron, manganese, zinc, molybdenum, copper, and chlorine.

Soils can be roughly divided into three main types on the basis of their usefulness horticulturally, but many areas contain a mixture.

Clays

Clays, in which the particles are very fine, are called in horticulture heavy soils, because it is difficult to turn them over with a spade. They can be very fertile but tend to be lacking in good drainage, holding their water closely adhered to the soil particles; therefore, they cannot be worked when wet, and under pressure they tend to compact tightly, driving out the air. During drought they tend to become hard and even to develop large cracks so that they cannot be worked satisfactorily. Clay soils can be lightened with as much humus as can be dug into them. Humus may be any decayed organic matter, such as farmyard manure, leaf mold, or compost made from kitchen scraps and grass clippings.

Sands and Gravels

Sands and gravels are opposite in properties to clay. The soil particles are large, and the soils are called light because they are easy to work and turn in nearly all weather. Since their water-holding capacity is very low, however, they tend to dry out quickly. They are "hungry" soils requiring great quantities of manures, humus, and fertilizers to keep them prolific.

Peats and Heaths

Peats and heaths are usually very acid and ill-drained. They result where conditions have prevented the complete breakdown of old vegetable matter into humus, generally because of poor aeration and surplus acid bog water. Much peat is derived from the decaying roots of sphagnum moss, useful for mulching in the garden. A heath soil is generally less fertile, consisting of a large mixture of sand with the peat and tending to be very low in mineral content and in water-retaining capacity.

The ideal garden soil is a medium loam consisting of a mixture of clay and sand, fairly rich in humus and easily worked, and not forming large clods when dry. The consistency of the soil is important, for a porous, properly tilled soil provides a medium through which roots can penetrate readily and rapidly. Another factor of importance in soils is the degree of acidity or alkalinity. Soil alkalinity is usually derived from free calcium carbonate or a similar alkaline salt. Soil reaction can be modified. It may be made more alkaline by adding one of the organic salts, of which calcium is best, in the form of lime. Acidity may be increased by adding hydrogen, in the form of sulfur compounds such as ammonia sulfate or superphosphate.

Feeding: Fertilizing and Watering

Maximum return can be obtained only from soil with an ample supply of elements necessary for plantgrowth, combined with sufficient moisture to enable them to be dissolved and absorbed through the plant hairs.

Treatment with farmyard manure or garden compost can supply the majority of these requirements. Because manure and compost are scarce in urban areas it is often necessary to use mineral fertilizers as well as organics. The soil is such a complex substance that all fertilizers must be applied in moderation and in balance with each other according to the deficiencies of the soil and the requirements of the particular crop. Different crops have different fertilizer needs. Manures are generally best dug into the ground in autumn in a temperate climate but also may be used as mulches in spring to control weeds. A mulch is a surface layer of organic matter that helps the several needs of feeding, conserving moisture, and controlling weeds. Black polyethylene sheeting is now widely used for all the mulching functions except feeding.

Watering of newly placed plants and of all plants during periods of drought is an essential gardening chore. Deep and thorough watering—not simply sprinkling the soil surface—can result in greatly improved growth. Water is essential in itself, but it also makes minerals available to plants in solution, the only form usable by plants. About one inch of water applied each week to the soil surface will percolate down about six inches; this is a minimal subsistence amount for many herbaceous garden plants, and small trees and shrubs require more. Proper watering once a week encourages deep penetration of roots, which in turn enables plants to survive dry surface conditions.

Drainage

Drainage is the other important side of water management. All plants need water but the amount needed varies, and if plants are forced to absorb more than they need, a form of drowning occurs. The symptoms are most easily seen in overwatered pot plants but are also visible to an experienced eye in badly drained corners of a garden. Roots require air as well as water and depend on subsurface water to bring the necessary oxygen. In large private gardens and in commercial gardens, buried earthenware piping is commonly used. In smaller gardens drainage can be readily achieved by the use of sumps, that is, holes dug to a depth of about four feet in affected places. The bottom half of the sump is filled with stones, through which excess water drains. Such measures may greatly improve the potential of a garden and the workability of its soil.

Protecting Plants

Most plants have a precise level of tolerance to cold, below which they are killed. Many plants from tropical or subtropical regions cannot survive frost and are killed by temperatures below 32 °F (0 °C). These are called frost-tender. Others, called half-hardy, can withstand a few degrees of frost. Fortunately, many of the best garden plants are completely hardy, a quality often encouraged by careful breeding, and will withstand any low temperatures likely to be reached in temperate regions.

Various measures can be taken to give frost protection, from the simple ones appropriate for smaller gardens to the elaborate coverings used to protect valuable horticultural crops. Removing weeds that shade the soil increases the amount of heat stored during the day. Well-drained soil is less susceptible. Any shield against wind in frosty weather enhances survival capability. The simplest form of protection is a wrapping to keep warmer air around the plant. This can be a mulch (leaves, soil, ashes) placed over the crown of a slightly tender plant in winter or a shield of sacking for leaf-shedding plants (not as desirable for evergreens, which utilize their leaves all the year).

Glass structures such as greenhouses or outdoor frames can provide additional protection for tender plants. Such structures can be heated and the temperature regulated by a thermostat to any required degree. Thus, in temperate regions, orchids and other tropical plants can be grown so that they flowerthroughout the winter, many being forced to flower earlier than their normal season by the higher temperature. Greenhouses are divided by gardeners into four rough categories: (1) The cold house, in which there is no supplementary heating and which is suitable only for plants that will not be killed by a few degrees of frost (such as alpines or potted bulbous plants). The combination of heat from the sun and protection from wind will keep such a house appreciably warmer than the temperature outside. (2) The coolhouse, in which the minimum temperature is kept to 45 °F (7 °C). Most amateurs' greenhouses fall into this class, and a very large range of plants can be grown in them. (3) The intermediate house, in which the minimum temperature is kept at 55 or 60 °F (13 or 16 °C) and which is suitable for a wide range of orchids. (4) The hothouse, or stove house, in which the minimum temperature is kept above 60 °F (16 °C) and in which tropical plants such as anthuriums and cattleyas (a genus of the orchid family) can be grown.

Training and Pruning

Training, the orienting of the plant in space, is achieved by techniques that direct the shape, size, and direction of plant growth. It may be accomplished by use of supports to which plants can be bent, twisted, or fastened. Pruning, the judicious cutting away of plant parts, is performed for other purposes: to contain size, to encourage fruiting in orchard trees, or to improve the appearance of ornamental trees and shrubs. It is one of the most important horticultural arts.

Where trees and shrubs are left to grow naturally, they often become much too large for their space in the garden. Also they may grow lanky and misshapen and have much dead growth. Where a branch or shoot is cut, it will often be induced to make a number of young shoots from below the cut, and these are likely to flower more freely than the older branches. Fruit trees in particular when pruned annually often give fruit of finer quality, larger in size, freer of disease, and of better colour. The two basic pruning cuts are known as heading back and thinning out. Heading back consists of cutting back the terminal portion of a branch to a bud; thinning out is the complete removal of a branch to a lateral or main trunk. Heading back, usually followed by the stimulation of lateral budbreak below the cut, produces a bushy, compact plant, suitable for a hedgerow, and it is often used to rejuvenate shrubs that have become too large or that flower poorly. Thinning out, which encourages longer growth of the remaining terminals by reducing lateral branches, tends to open up the plant, producing a longer plant. In general, pruning, started when the plant is young, obviates the need for drastic and risky remedial pruning later of a large, old, or misshapen bush or tree.

Particular spatial arrangements may increase light utilization, facilitate harvesting or disease control, or improve productivity and quality. Thus, training and pruning form an essential part of fruit growing throughout the life of the plant. Special attention is given in the formative years to obtain desired shape and structure. The key to training is the point on the main stem from which branches form. In the central-leader system of training, the trunk forms a central axis with branches distributed laterally up and down and around the stem. In the open-centre or vase system, the main stem is terminated and growth forced through a number of branches originating close to the upper end of the trunk. An intermediate system is called the modified-leader system. In espalier systems plants are trained to grow flat along a wire or trellis. Properly executed espaliers are extremely attractive as ornamentals. Espaliers in combination with dwarfing rootstocks allow high-density orchards that are very productive on a per-unit-area basis, with the fruit close to the ground for easy harvest. Extensive pruning is required annually to maintain the system.

There are a number of physiological responses to training and pruning. Orientation of the plant may have a marked effect on growth and fruiting. Thus fruit trees planted on an inclined angle become dwarfed and flower earlier; training branches in a horizontal position produces the same effect. This effect is achieved naturally when a heavy fruit load bends a limb down. The main effects of pruning are achieved by altering the root–shoot balance. Thus an explosion of vegetative growth normally occurs after extensive shoot pruning. Severely pruned plants, especially if they are in the juvenile stage of growth, tend to remain vegetative. Similarly the slowdown of vegetative growth by root pruning encourages flowering.

The training of plants to grow in unnatural shapes for ornamental purposes is called topiary. In Roman and Renaissance times, when ingenious topiary was in high fashion, plants were trained to unusual and fantastic shapes such as beasts, ships, and building facades. Though more modestly, hedges and shrubs are still trained to geometric shapes in formal gardens.

Another extreme form of training is the Japanese art of bonsai, the creation of dwarfed potted trees by a combination of pruning (both roots and tops) and restricted nutrition. Living trees more than 100 years old and only a few feet high are grown in special containers arranged to resemble the natural landscape.

Propagation

New plants are produced either from seed or by the techniques of division, taking cuttings, grafting, budding, or layering. For the ordinary gardener, propagation is a relatively simple but interesting process normally used for economic provision of more versions of favourite plants, as part of exchanges with other gardeners, or as a wise precaution against winter losses.

Propagation by cuttings is the most common practice. Young shoots of the current season are usually the most successful at rooting. Roses are usually propagated by budding, in which a bud from the rose desired is inserted in rootstock just above ground level. Fruit trees are usually propagated by layering, in which a young shoot is pegged down in the ground with the end twisted upward almost at right angles; the lower side of the wood just before the twist is wounded so as to induce rooting. When this has taken place, the layer is severed from the parent.

Control of Weeds

Controlling weeds is a basic, and probably the most arduous, factor of cultivation and has been carried on from the time the earliest nomads settled down to an agricultural life. It has always been necessary to free the chosen crops of competition from other plants. For smaller weeds hoeing is practicable. The weeds are cut off by the action of the hoe and left to wither on the surface. Hand weeding, by pulling out individual weeds, is often necessary in gardens, particularly the rock garden, in seed boxes, and in the herbaceous border or among annuals. Chemical and biological control of weeds developed greatly after World War II and has made much mechanical cultivation unnecessary.

Control of Pests and Diseases

Damage to plants is most often caused by pests such as insects, mites, eelworms, and other small creatures but may also be caused by mammals such as deer, rabbits, and mice. Damage by disease is that caused by fungi, bacteria, and viruses.

Prevention is generally better than cure, and constant vigilance is necessary to prevent a pest infestation or a disease outbreak. Control can be obtained by the use of chemical sprays, dusts, and fumigants, but some of these are so potent that they should be used only by the experienced operator. Considerable evidence is available regarding the possible harmful long-term effects on the biological chain of excessive use of some of these noxious chemicals, particularly the hydrocarbons. Some control can be obtained through good garden practices: clearing up all dead and diseased material and burning it; pruning and thinning so that a reasonable circulation of air is obtained through the plants; and crop rotation. Some control may also be obtained through natural biological predators. The breeding of plants immune to certain pests and diseases is also a valuable means of control.

Mechanical Aids

Mechanical devices to aid the gardener include tillers, lawn mowers, hedge cutters, sprinklers, and a variety of more esoteric equipment that has made gardening an easier pursuit. Such machines are not a substitute for good judgment and technique in the garden, however, nor will they give anyone a completely labour-free garden. They do enable a considerably larger area to be cultivated and maintained than if all labour is performed by hand.

Landscape Gardening

Landscape gardening is an aesthetic branch of Horticulture which deals with planting of ornamental plants in such a way that it creates a picturesque effect. It is a very fascinating and interesting subject. There are several definitions and expressions to define this subject. landscape is the appearance of that portion of land which the eye can view at once and landscape gardening is the art of laying grounds so as to produce the effect of a picturesque landscape. Landscape gardening can be defined as the decoration of a tract of land with plants and other garden materials so as to produce a picturesque and naturalistic effect in a limited space. So landscape may or may not include plants. According to Bailey, Landscape gardening is the application of garden forms, methods and materials to the improvements of the landscape and the landscape in this connection is any area large or small on which it is desirable to develop a view or design.

Landscape gardening can also be defined as the beautification of a tract of land having a house or other object of interest on it. It is done with a view to create a natural scene by the planting of lawn, trees and shrubs. Landscape gardening is both an art and science of the establishment of a ground in such a way that it gives an effect of a natural landscape. It can be also defined as the imitation of nature in the garden. It can also be defined as improving of total living environment for the people. The expression of landscape may be gay, bold, retired, quiet, etc. This expression will conform to the place and the purpose. It should be a picture and not a collection of interesting objects.

Since the landscape gardening is the making pictures on the ground with plant and other material, landscape designer should be proficient in art, ornamental gardening, ecology and physiology. He should be an architect and engineer to appreciate the relationship between plant form, colours and buildings.

Natural Elements of Landscape

Different types of landscape depending upon prevailing geographical and agroclimatic conditions characterize Earth. There are mountains, hills, glens, valleys, seas, rivers, forests, plains, deserts, lakes, swamps, streams, etc. which comprise major part of natural landscape. At certain points, there is harmony between natural elements like ground forms, vegetation and even animal life. The landscape of such place is beautiful and conveys the feeling or mood of the landscape character like exhilaration, sadness, ceriness or awe. There are many qualities of natural landscape beauty like the picturesque; the ethereal, the serene, the delicate, the idyllic, the graceful, the majestic, the bold etc.

Man has copied the natural elements for improving landscape around him and converted certain areas in the form of garden for his pleasure.

Landscaping Principles

General principles of landscaping are as follows:

- The ideal landscape garden is like ideal landscape painting which expresses some single thought or feelings. Its expression may be gay, bold, retired, quiet, etc.
- Beauty and utility should be harmoniously combined.
- Area should be divided into different parts and plan should be conceived for each area. Overall plan should be such that the observer catches the entire effect and purpose of the plan without stopping to analyse its parts.
- Simplicity of design should be aimed at in the execution of the plan.
- An ideal landscape should have open space.
- Let the garden and building merge into each other. There should not be stopping abruptly particularly in front of building. The view of garden from the windows and doors should be very attractive. Planting around the building, climbers against wall and on the porch, decoration of verandah and rooms with attractive foliage, flowering plants, hanging baskets serve to unique the building with garden. Every part of the compound should be planned in such a way that it gives surprising effect to visitor.
- Over crowding of plants and objects should be avoided.

Factors affecting the Landscape Design

There are several factors which affect the making of suitable design for particular site. These factors are:

- Human choice: Man's ultimate desire is to make his living pleasurable and surroundings confortable. His dominance in making designs and selection of plant material is very well evident. Therefore, different styles of gardening have come into existence.
- Site: This is an important factor and according to site, suitable design is made. In formal style gardening, the site is selected according to plan. Topography of the site also affects the design.
- Views: Distant views of mountains, hills, woods, valley, etc. are preferred from the place of garden.
- Heritage: One inherits the knowledge of botany and aesthetic sense and uses accordingly. Our rich heritage teaches us to use flowers and fragrant trees to improve the surroundings.
- Climate: The climate of particular place affects the selection of plant material accordingly. Ideally suited plant material according to climate should be selected.
- Soil: According to characteristics of soil types, suitable plants should be selected.

Garden Styles

Garden styles have been changed from time to time with the new ideas and necessities. Broadly, the styles of gardening are grouped into three categories i.e. Formal style, Informal style and Free style of gardening:

- Formal style: Main features of this style of gardening are: First plan is made on the paper and then land is selected accordingly. Plan is symmetrical. These types of gardens are of geometric design i.e. squarish or rectangular. Therefore, the roads are cut at right angle. It has some sort of enclosure. Flower beds are also of geometric shapes. The arrangements of tree and shrubs are necessarily geometrical and kept in shape by trimming and training. Other features like fountains, water pools, cascades, etc. are used for further attraction. The examples of such style of gardening are Persian gardens and Mogul gardens.

- Informal Style: This style reflects naturalistic effect of total view and represents natural beauty. This style is just contrast of above formal style. In this, plan is asymmetrical and according to the land available for making garden. Roads, paths are made curvaceous and bending. Water bodies are made of irregular shapes. Hillocks are made to create natural mountainous scenery. Flower beds are made of irregular shapes suiting to surroundings. Plants are allowed to grow in natural form and instead of trimming, annual pruning is done. Japanese gardens are the best example of this style of gardening.

- Free style of gardening: This style combines the good points of both formal and informal style of gardening. Rose garden of Ludhiana is an example of this style of gardening.

Art Principles of Landscape

Landscaping is making of pictures with plant material and, hence, its principles are same as those of art. They are as follows:

- Rhythm

Repetition of same object at equidistance is called rhythm. It can be created through the shapes, progression of sizes or a continuous line movement, rhythm creates movement to the eye. In gardens, generally trees of single species of equal height and shape are planted to create this effect. In Mogul gardens, fountains and water canals have also been extensively used to create such effect. Now-a-days other objects like lights are also used to create the effect of rhythm.

- Balance

It is very important to maintain the balance on both sides of the central line. The principle involved in making balance of see-saw game can help in understanding this. Equal weights can be balanced only when they are equidistant from the centre. If weights are unequal, the heavier must move towards centre for making balance. The balance may be formal, informal or symmetrical types. Imbalance will look lopsided and will distract the attention. In making the balance with the plants, their form, colour, texture etc. are kept in view.

- Accent or emphasis

The accent or emphasis is created in the gardens to avoid the monotonous view. It is the method to stress the most important thing. This also serves as the centre of attraction. Mostly unusual objects like tall fountain, tree, statue etc. are used to create the effect of accent or emphasis. In English gardens, statues have been used extensively to create such effects.

- Contrast

This principle is most useful in emphasising the best features of an object. It can be very easily understood by following contrast colour theory. Against green background, a fleck of scarlet colour will make a contrast and will make scarlet colour prominent. In nature this is very common. Other contrast colour can also be used. Similarly, weeping growth habit against upright growth, dwarf against tall, rough texture against soft texture etc, are some of the examples which can be followed. It is also very important that one of the two contrasting objects must clearly dominate each other. In this way, one becomes feature whereas other acts as supporting background. The contrasting elements of equal power may create visual tensions.

- Proportion

It is the relation of one thing to another in magnitude. When two or more objects are put together the proportions are established. In a landscape design, space provided for lawn, paths, herbaceous borders, shrubbery border, trees, buildings and other garden objects should be in a right proportion. It will create harmonious effect and look better. Such effects can be noticed in Persian and Mogul gardens. Out of proportion allotment of area in garden will distract the attention. Proportion helps in space organization.

- Harmony

It is an overall effect of various features, styles, and colour schemes of the total scene. The degree of harmony or unity of various elements of landscape is a measure induced in us and is called as beauty. Therefore, the beauty can be defined as the evident relationship of all parts of a thing observed. When different parts of landscape are correctly placed in right way, produces a harmonious effect. Such landscapes create picturesque effect and appeals to visitors. On the contrary, the absence of harmony or lack of unity is ugliness.

Some Important Landscape Gardening Terms

Axis: It is an imaginary line, which divides garden into two parts. This also connects two or more points. It is presented in the form of a path, line of fountains or trees, etc. If this axis divides garden into two equal parts, it is called central axis. In formal style, axis is central whereas in informal style, it is oblique. This axis controls the movement in the garden from the entrance to the terminal. An axis in garden is directional, orderly or dominating.

Symmetrical plan: In this plan, the different objects are in equilibrium about a central point or on either side of an axis. Symmetrical plan or formal plan is synonymous of beauty and is pleasant and handsome. This is because the symmetry is to be associated with plan clarity, rhythm, balance, unity, etc. Symmetrical plan being precise and disciplined, it requires precision in detail and maintenance and bold in concept.

Dynamic Symmetry: In such symmetrical plan, each pole generates its own magnetic field and between these two fields there is a field of dynamic tension.

Asymmetrical Plan: In such plans, there is absence of symmetry on both the sides of axis but balance, unity and harmony are maintained.

Circulation in Landscaping: In landscape gardening it means a pathway from entrance to terminal point. Circulation varies with the style of gardening and topography. The more of circulation patterns, it has more points of views and attraction. Common circulation patterns are depicted.

Vista: It is a three dimensional confined view of terminal building or dominant element of feature. It may be natural or man-made. Natural vistas are very common around the lofty mountains and snowy peaks. Overall effect of vista may be of its characteristics. It may be calm or induce motion. There are three different parts of a vista viz.

- It should be subject to a close control.
- It should have a viewing station to see object or objects.
- It should have intermediate ground. These three should be satisfactorily united and thus result into an effect of totality.

Container Gardening

Container gardening or pot gardening is the practice of growing plants, including edible plants, exclusively in containers instead of planting them in the ground. A container in gardening is a small, enclosed and usually portable object used for displaying live flowers or plants. It may take the form of a pot, box, tub, pot, basket, tin, barrel or hanging basket.

Methods

Pots, traditionally made of terracotta but now more commonly plastic, and window boxes have been the most commonly seen. Small pots are commonly called flowerpots. In some cases, this method of growing is used for ornamental purposes. This method is also useful in areas where the soil or climate is unsuitable for the plant or crop in question. Using a container is also generally necessary for houseplants. Limited growing space, or growing space that is paved over, can also make this option appealing to the gardener. Additionally, this method is popular for urban horticulture on balconies of apartments and condominiums where gardeners lack the access to the ground for a traditional garden.

Planting

Containers range from simple plastic pots, teacups to complex automatic-watering irrigation systems. This flexibility in design is another reason container gardening is popular with growers. They can be found on porches, front steps, and in urban locations, on rooftops. Sub-irrigated planters (SIP) are a type of container that may be used in container gardens. Potting material must be loose and allow drainage to offer proper aeration for roots to breathe, preventing root rot.

Re-potting

Root-bound.

Re-potting is the action of placing an already potted plant into a larger or smaller pot. A pot that fits a plant's root system better is normally used. Plants are usually re-potted according to the size of their root system. Most plants need to be re-potted every few years because they become "pot-bound" or "root-bound". Plants' roots can sense its surroundings, including the size of the pot it is in, and increasing the pot size allows plant size to increase proportionally.

Advantages

Many types of plants are suitable for the container, including decorative flowers, herbs, cacti, vegetables, and small trees. There are many advantages to growing plants in containers, namely:

- Less risk of soil-borne disease.
- Virtually eliminate weed problems.
- Mobile plants gives more control over moisture, sunlight & temperature.
- Great addition to the interior of the house.

Sustainable Gardening

Sustainable gardening includes the more specific sustainable landscapes, sustainable landscape design, sustainable landscaping, sustainable landscape architecture, resulting in sustainable sites. It comprises a disparate group of horticultural interests that can share the aims and objectives associated with the international post-1980s sustainable development and sustainability programs developed to address the fact that humans are now using natural biophysical resources faster than they can be replenished by nature.

Included within this compass are those home gardeners, and members of the landscape and nursery industries, and municipal authorities, that integrate environmental, social, and economic factors to create a more sustainable future.

Organic gardening and the use of native plants are integral to sustainable gardening.

Principles and Concepts

Managing global biophysical cycles and ecosystem services for the benefit of humans, other organisms and future generations has now become a global human responsibility. The method of applying sustainability to gardens, landscapes and sites is still under development and varies somewhat according to the context under consideration. However, there are a number of basic and common underlying biological and operational principles and practices in the sustainable sites literature.

Biological Principles

Sustainable management of man-made landscapes emulates the natural processes that sustain the biosphere and its ecosystems. First and foremost is the harnessing the energy of the Sun and the cycling of materials thereby minimising waste and energy use.

Running within, and dependent on, the natural economy there is the production and consumption of goods and services in the "human economy" which has now significantly altered, in a detrimental way, natural biogeochemical cycles.

Native Plants

The use of native plants in a garden or landscape can both preserve and protect natural ecosystems, and reduce the amount of care and energy required to maintain a healthy garden or landscape. Native plants are adapted to the local climate and geology, and often require less maintenance than exotic species. Native plants also support populations of native birds, insects, and other animals that they coevolved with, thus promoting a healthy community of organisms.

Plants in a garden or maintained landscape often form a source population from which plants can colonize new areas. Avoiding the use of invasive species helps to prevent such plants from establishing new populations. Similarly, the use of native species can provide a valuable source to help these plants colonise new areas.

Some non-native species can form an ecological trap in which native species are lured into an environment that appears attractive but is poorly suited to them.

However, in Britain research by the University of Sheffield as part of the BUGS project (Biodiversity in Urban Gardens in Sheffield) has revealed that for many invertebrates - the majority of wild animals in most gardens - it is not just native plants which can sustain them. The findings were published in popular form in Ken Thompson's book 'No Nettles Required: The truth about wildlife gardening'.

Operational Principles

Enhancement of ecosystem services is encouraged throughout the lifecycle of any site by providing clear design, construction, (operations), and management criteria. To be sustainable over the long term requires environmental, social and economic demands are integrated to provide intergenerational equity by providing regenerative sustainable systems. Operational guidelines will link to and supplement existing guidelines for the built environment (supplementing existing green building and landscape guidelines), the wider environment, and they will include metrics (benchmarks, audits, criteria, indexes etc.) that give some measure of sustainability (a rating system) by clarifying what is sustainable or not sustainable or, more likely, what is more or less sustainable.

Scale

Impacts of a site can be assessed and measured over any spatio-temporal scale or context.

Direct and Indirect Environmental Impact

Impacts of a site may be direct by having direct measurable impacts on biodiversity and ecology at the site itself or indirect when impacts occur away from the site.

Site Principles

Compost heap at the Royal Botanic Gardens, Kew.

The following are some site principles for sustainable gardening:

- Do no harm.
- Use the Precautionary principle.
- Design with nature and culture.
- Use a decision-making hierarchy of preservation, conservation, and regeneration.
- Provide regenerative systems as intergenerational equity.
- Support a living process.
- Use a system thinking approach.

- Use a collaborative and ethical approach.
- Maintain integrity in leadership and research.
- Foster environmental stewardship.

Measuring Site Sustainability

One major feature distinguishing the approach of sustainable gardens, landscapes and sites from other similar enterprises is the quantification of site sustainability by establishing performance benchmarks. Because sustainability is such a broad and inclusive concept the environmental impacts of sites can be categorised in numerous ways depending on the purpose for which the figures are required. The process can include minimising negative environmental impacts and maximising positive impacts. As currently applied the environment is usually given priority over social and economic factors which may be added in or regarded as an inevitable and integral part of the management process. A home gardener is likely to use simpler metrics than a professional landscaper or ecologist.

Three methodologies for measuring site sustainability include BREEAM developed by the BRE organisation in the UK, Leed, developed in America and the Oxford 360 degree sustainability Index used in Oxford Park and developed by the Oxford Sustainable Group in Scandinavia.

The *Sustainable Sites Initiative* is producing recommendations for the American Landscape Industry. The standards and guidelines finally adopted will lead to a uniform national standard, which does not currently exist. Sustainable Sites is currently in the pilot program stage, and will formally introduce its first rating system by 2013. The U.S. Green Building Council supports the project and plans to adopt the Sustainable Sites metrics into future versions of its Leadership in Energy and Environmental Design Green Building Rating System. Sites are rated according to their impact on ecosystem services: The following ecosystem services have been identified by the study group:

- Local climate regulation.
- Air and water cleansing.
- Water supply and regulation.
- Erosion and sediment control.
- Hazard mitigation.
- Pollination.
- Habitat functions.
- Waste decomposition and treatment.
- Global climate regulation.
- Human health and well-being benefits.
- Food and renewable non-food products.
- Cultural benefits.

Inputs	Outputs
• Fossil fuels • Embodied energy and water • Compost • Mulch • Ecology & biodiversity • Fertilizer • Hard landscape materials • Equipment • Products	• Energy & water • Food • Green waste • Ecology & biodiversity • Chemicals • Old hard landscape materials • Old equipment • Old products

Constraints

Any kind of auditing or benchmarking will depend on the selection and weighting of the metrics chosen; the depth and detail of analysis required; the purpose for which the figures are required; and the environmental circumstances of the particular site.

Bonsai Cultivation and Care

Bonsai cultivation and care involves the long-term cultivation of small trees in containers, called bonsai in the Japanese tradition of this art form. Trees are difficult to cultivate in containers, which restrict root growth, nutrition uptake, and resources for transpiration (primarily soil moisture). In addition to the root constraints of containers, bonsai trunks, branches, and foliage are extensively shaped and manipulated to meet aesthetic goals. Specialized tools and techniques are used to protect the health and vigor of the subject tree. Over time, the artistic manipulation of small trees in containers has led to a number of cultivation and care approaches that successfully meet the practical and the artistic requirements of bonsai and similar traditions.

The term bonsai is generally used in English as an umbrella term for all miniature trees in containers or pots. In this bonsai should be understood to include any container-grown tree that is regularly styled or shaped, not just one being maintained in the Japanese bonsai tradition.

Bonsai can be created from nearly any perennial woody-stemmed tree or shrub species which produces true branches and remains small through pot confinement with crown and root pruning. Some species are popular as bonsai material because they have characteristics, such as small leaves or needles, that make them appropriate for the compact visual scope of bonsai. Bonsai cultivation techniques are different from other tree cultivation techniques in allowing mature (though miniature) trees to grow in small containers, to survive with extremely restricted root and canopy structures, and to support comprehensive, repeated styling manipulations.

Sources of Bonsai Material

All bonsai start with a specimen of source material, a plant that the grower wishes to train into bonsai form. Bonsai practice is an unusual form of plant cultivation in that growth from seeds is rarely used to obtain source material. To display the characteristic aged appearance of a bonsai within a reasonable time, the source plant is often partially grown or mature stock. A specimen may be selected specifically for bonsai aesthetic characteristics it already possesses, such as great natural age for a specimen collected in the wild, or a tapered, scar-free trunk from a nursery specimen. Alternatively, it may be selected for non-aesthetic reasons, such as known hardiness for the grower's local climate or low cost (as in the case of collected materials).

Propagation

Plant cuttings can be rooted and grown as potential bonsai.

While any form of plant propagation could generate bonsai material, a few techniques are favored because they can quickly produce a relatively mature trunk with well-placed branches.

Cuttings: In taking a cutting, part of a growing plant is cut off and placed in a growing medium to develop roots. If the part that is cut off is fairly thick, like a mature branch, it can be grown into an aged-looking bonsai more quickly than can a seed. Unfortunately, thinner and younger cuttings tend to strike roots more easily than thicker or more mature ones. In bonsai propagation, cuttings usually provide source material to be grown for some time before training.

Layering: Layering is a technique in which rooting is encouraged from part of a plant, usually a branch, while it is still attached to the parent plant. After rooting, the branch is removed from the parent and grown as an independent entity. For bonsai, both ground layering and air layering can create a potential bonsai, by transforming a mature branch into the trunk of a new tree. The point at which rooting is encouraged can be close to the location of side branches, so the resulting rooted tree can immediately have a thick trunk and low branches, characteristics that complement bonsai aesthetics.

Commercial Bonsai Growers

Commercial bonsai growers may use any of the other means of obtaining starter bonsai material, from seed propagation to collecting expeditions, but they generally sell mature specimens that display bonsai aesthetic qualities already. The grower trains the source specimens to a greater or lesser extent before sale, and the trees may be ready for display as soon as they are bought. Those who purchase commercially grown bonsai face some challenges, however, particularly of buying from another country. If the purchaser's local climate does not closely match the climate in which the bonsai was created, the plant will have difficulties surviving and thriving. As well, importing living plant material from a foreign source is often closely controlled by import regulations and may require a license or other special import arrangement on the buyer's part. If a local commercial bonsai grower does not exist, buying from a distant one may be unsatisfactory.

Nursery Stock

A plant nursery is an agricultural operation where (non-bonsai) plants are propagated and grown to usable size. Nursery stock may be available directly from the nursery, or may be sold in a garden centre or similar resale establishment. Nursery stock is usually young but fully viable, and is often potted with sufficient soil to allow plants to survive a season or two before being transplanted into a more permanent location. Because the nursery tree is already pot-conditioned, it can be worked on as a bonsai immediately. The large number of plants that can be viewed in a single visit to a nursery or garden centre allows the buyer to identify plants with better-than-average bonsai characteristics. According to Peter Adams, a nursery visit "offers the opportunity to choose an instant trunk". One issue with nursery stock is that many specimens are shaped into popular forms, such as the standard or half-standard forms, with several feet of clear trunk rising from the roots. Without branches low on the trunk, it is difficult for a source specimen to be trained as bonsai.

Collecting

Collecting bonsai consists of finding suitable bonsai material in its original wild situation, successfully moving it, and replanting it in a container for development as bonsai. Collecting may involve wild materials from naturally treed areas, or cultivated specimens found growing in yards and gardens. For example, mature landscape plants being discarded from a building site can provide excellent material for bonsai. Hedgerow trees, grown for many years but continually trimmed to hedge height, provide heavy, gnarled trunks for bonsai collectors. In locations close to a tree line (the line beyond which trees do not grow, whether due to altitude, temperature, soil moisture, or other conditions), aged and naturally dwarfed survivors can be found.

The main benefit of collecting bonsai specimens is that collected materials can be mature, and will display the natural marks and forms of age, which makes them more suitable for bonsai development than the young plants obtained through nurseries. Low cost is another potential benefit, with a tree harvest license often being more economical than purchase of nursery trees. Some of the difficulties of collecting include finding suitable specimens, getting permission to remove them, and the challenges of keeping a mature tree alive while transplanting it to a bonsai pot.

Styling Techniques

This juniper makes extensive use of both jin (deadwood branches) and shari (trunk deadwood).

Bonsai are carefully styled to maintain miniaturization, to suggest age, and to meet the artist's aesthetic goals. Tree styling also occurs in a larger scale in other practices like topiary and niwaki. In bonsai, however, the artist has close control over every feature of the tree, because it is small and (in its container) easily moved and worked on. The greater scale of full-sized trees means that styling them may be restricted to pruning and shaping the exterior volume once per growing season, never pruning within the canopy nor bending and forming individual branches. In contrast, in a bonsai being prepared for display, each leaf or needle may be subject to decision regarding pruning or retention, and every branch and twig may be formed and wired into place each year. Given these differences in scope and purpose, bonsai styling uses a number of styling techniques either unique to bonsai or (if used in other forms of plant cultivation) applied in ways particularly suitable to meet the goals of bonsai development.

Leaf Trimming

This technique involves selective removal of leaves (for most varieties of deciduous tree) or needles (for coniferous trees and some others) from a bonsai's branches. A common aesthetic technique in bonsai design is to expose the tree's branches below groups of leaves or needles (sometimes called "pads") by removing downward-growing material. In many species, particularly coniferous ones, this means that leaves or needles projecting below their branches must be trimmed off. For some coniferous varieties, such as spruce, branches carry needles from the trunk to the tip and many of these needles may be trimmed to expose the branch shape and bark. Needle and bud trimming can also be used in coniferous trees to force back-budding on old wood, which may not occur naturally in many conifers. Along with pruning, leaf trimming is the most common activity used for bonsai development and maintenance, and the one that occurs most frequently during the year.

Pruning

The small size of the tree and some dwarfing of foliage result from pruning the trunk, branches, and roots. Pruning is often the first step in transforming a collected plant specimen into a candidate for bonsai. The top part of the trunk may be removed to make the tree more compact. Major and minor branches that conflict with the designer's plan will be removed completely, and others may be shortened to fit within the planned design. Pruning later in the bonsai's life is generally less severe, and may be done for purposes like increasing branch ramification or encouraging growth of non-pruned branches. Although pruning is an important and common bonsai practice, it must be done with care, as improper pruning can weaken or kill trees. Careful pruning throughout the tree's life is necessary, however, to maintain a bonsai's basic design, which can otherwise disappear behind the uncontrolled natural growth of branches and leaves.

Wiring

Extensive wiring can be seen on this bonsai specimen.

Wrapping copper or aluminium wire around branches and trunks allows the bonsai designer to create the desired general form and make detailed branch and leaf placements. When wire is used on new branches or shoots, it holds the branches in place until they lignify (convert into wood). The time required is usually 6–9 months or one growing season for deciduous trees, but can be several years for conifers like pines and spruce, which maintain their branch flexibility through multiple growing seasons. Wires are also used to connect a branch to another object (e.g., another branch, the pot itself) so that tightening the wire applies force to the branch. Some species do not lignify strongly, and some specimens' branches are too stiff or brittle to be bent easily. These cases are not conducive to wiring, and shaping them is accomplished primarily through pruning.

Clamping

For larger specimens, or species with stiffer wood, bonsai artists also use mechanical devices for shaping trunks and branches. The most common are screw-based clamps, which can straighten or bend a part of the bonsai using much greater force than wiring can supply. To prevent damage to the tree, the clamps are tightened a little at a time and make their changes over a period of months or years.

Grafting

In this technique, new growing material (typically a bud, branch, or root) is introduced to a prepared area under the bark of the tree. There are two major purposes for grafting in bonsai. First, a number of favorite species do not thrive as bonsai on their natural root stock and their trunks are often grafted onto hardier root stock. Examples include Japanese red maple and Japanese black pine. Second, grafting allows the bonsai artist to add branches (and sometimes roots) where they are needed to improve or complete a bonsai design. There are many applicable grafting techniques, none unique to bonsai, including branch grafting, bud grafting, thread grafting, and others.

Defoliation

Short-term dwarfing of foliage can be accomplished in certain deciduous bonsai by partial or total defoliation of the plant partway through the growing season. Not all species can survive this technique. In defoliating a healthy tree of a suitable species, most or all of the leaves are removed by clipping partway along each leaf's petiole (the thin stem that connects a leaf to its branch). Petioles later dry up and drop off or are manually removed once dry. The tree responds by producing a fresh crop of leaves. The new leaves are generally much smaller than those from the first crop, sometimes as small as half the length and width. If the bonsai is shown at this time, the smaller leaves contribute greatly to the bonsai aesthetic of dwarfing. This change in leaf size is usually not permanent, and the leaves of the following spring will often be the normal size. Defoliation weakens the tree and should not be performed in two consecutive years.

Deadwood

Bonsai growers create or shape dead wood using techniques such as jin and shari to simulate age and maturity in a bonsai. Jin is the term used when the bark from an entire branch is removed to create the impression of a snag of deadwood. Shari denotes stripping bark from areas of the trunk to simulate natural scarring from a broken limb or lightning strike. In addition to stripping bark, deadwood techniques may also involve the use of tools to scar the deadwood or to raise its grain, and the application of chemicals (usually lime sulfur) to bleach and preserve the exposed deadwood.

Care

Small trees grown in containers, like bonsai, require specialized care. Unlike most houseplants, flowering shrubs, and other subjects of container gardening, tree species in the wild generally grow individual roots up to several meters long and root structures encompassing hundreds or thousands of liters of soil. In contrast, a typical bonsai container allows a fraction of a meter for root extension, and holds 2 to 10 liters of soil and root mass. Branch and leaf (or needle) growth in trees is also large-scale in nature. Wild trees typically grow 5 meters or taller when mature, while the largest bonsai rarely exceed 1 meter and most specimens are significantly smaller. These size differences affect maturation, transpiration, nutrition, pest resistance, and many other aspects of tree biology. Maintaining the long-term health of a tree in a container requires a number of specialized care techniques.

Growing Environment

Most bonsai species are trees and shrubs that must by nature grow outdoors. They require temperature, humidity, and sunlight conditions approximating their native climate year round. The skill of the grower can help bonsai from outside the local hardiness zone survive and even thrive, but doing so takes careful watering, shielding of selected bonsai from excessive sunlight or wind, and possibly protection from winter conditions (e.g., through the use of cold frames or winter greenhouses).

Common bonsai species (particularly those from the Japanese tradition) are temperate climate trees from hardiness zones 7 to 9, and require moderate temperatures, moderate humidity, and full sun in summer with a dormancy period in winter that may need be near freezing. They do not thrive indoors, where the light is generally too dim, and humidity often too low, for them to grow properly. Only during their dormant period can they safely be brought indoors, and even then the plants require cold temperatures, reduced watering, and lighting that approximates the number of hours the sun is visible. Raising the temperature or providing more hours of light than available from natural daylight can cause the bonsai to break dormancy, which often weakens or kills it.

Even for bonsai specimens that are native to the grower's location, outdoor cultivation requires specific cultivation practices to ensure successful long-term survival of the bonsai. The trees used in bonsai are constrained by the need to grow in a relatively small pot. This state greatly reduces the volume of roots and soil normally available to a freely grown tree, and brings the roots much closer to the surface of the soil than would occur in the wild. Trees in bonsai pots have much less access to water and to nutrients than they do natively, and physically confining roots changes their growth pattern and indirectly the growth pattern of the tree above the soil.

The grower has some control over the following environmental variables, and by controlling them effectively for individual specimens can ensure the health of native species grown as bonsai, and can cultivate some non-native species successfully.

- Watering: Different species of tree have roots with different tolerances for soil moisture. Some species tolerate continual wetness, while others are prone to rotting if the soil remains wet for long periods. A standard bonsai practice is to grow trees in a soil mixture that drains rapidly, so that roots are not allowed to be wet for long. To compensate for the relatively low water retention of the bonsai soil, water is applied frequently. The tree absorbs sufficient moisture for its needs while the water is passing through the soil, then the soil dries enough to reduce the chance of rotting. It is the grower's responsibility to ensure that watering occurs frequently enough to satisfy the bonsai with high watering requirements, while not waterlogging trees that use little water or have roots prone to rotting.
- Soil volume: Giving a bonsai a relatively large soil volume encourages the growth of roots, then corresponding growth of the rest of the tree. With a large amount of soil, the tree trunk extends in length and increases in diameter, existing branches increase in size and new branches appear, and the foliage expands in volume. The grower can move an outdoor bonsai from a pot to a training box or to open ground to stimulate this sort of growth. Replacing the tree in a bonsai pot will slow or halt the tree's growth, and may lead to die-back if the volume of foliage is too great for the limited root system to support. Managing the tree's available soil volume allows the grower to manage the overall size of the bonsai, and to increase vigor and growth when new branches are required for a planned styling.
- Temperature: Bonsai roots in pots are exposed to much greater variation in temperature than tree roots deep in the soil. For bonsai from native species, local temperatures do not generally harm the tree. But for bonsai from warmer native climates, the grower can increase the likelihood of successful cultivation either by insulating the tree from local winter

conditions, or by actively increasing the bonsai temperature during the cold season. For trees from climates slightly warmer than the local one, bonsai pots can be partially buried in the ground and can be covered with an insulating layer of mulch. For trees from significantly warmer climates, warmer temperatures can be maintained in a cold frame or greenhouse, so that a relatively tender tree is not exposed to temperatures lower than it can bear. This approach may also artificially extend the bonsai's growing season, affecting watering and fertilization schedules.

- Sunlight: Trees generally require a good deal of sun, and most bonsai need direct sunlight during the growing season to thrive. Some shade-tolerant species of bonsai cannot thrive with too much direct sunlight, however, and it is the grower's role to site the bonsai specimens to provide the correct lighting for each type. Most bonsai will be located in an area that gets several hours of direct daylight. Shade-tolerant bonsai can be placed behind barriers (walls, buildings), sited on shaded benches or stands, or shaded by netting to reduce the impact of direct sunlight.

Repotting

An uprooted bonsai, ready for repotting.

Bonsai are repotted and root-pruned at intervals dictated by the vigor and age of each tree. In the case of deciduous trees, this is done as the tree is leaving its dormant period, generally around springtime. Bonsai are often repotted while in development, and less often as they become more mature. This prevents them from becoming pot-bound and encourages the growth of new feeder roots, allowing the tree to absorb moisture more efficiently.

Specimens meant to be developed into bonsai are often placed in "growing boxes", which have a much larger volume of soil per plant than a bonsai pot does. These large boxes allow the roots to grow freely, increasing the vigor of the tree and helping the trunk and branches grow thicker. After using a grow box, the tree may be replanted in a more compact "training box" that helps to create a smaller, denser root mass which can be more easily moved into a final presentation pot.

Tools

Special tools are available for the maintenance of bonsai. The most common tool is the concave cutter (5th from left in picture), a tool designed to prune flush, without leaving a stub. Other tools include branch bending jacks, wire pliers and shears of different proportions for performing detail and rough shaping.

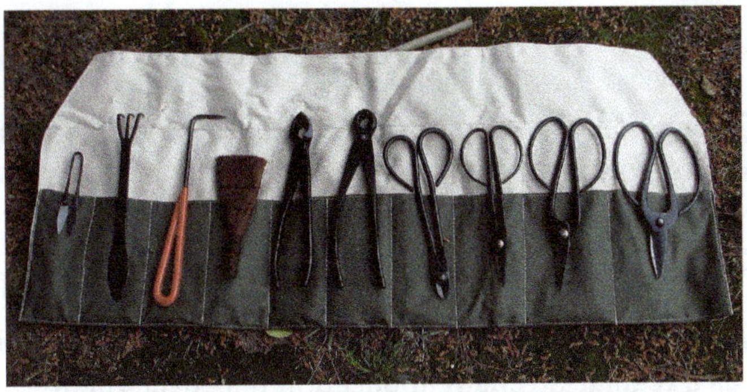

Set of bonsai tools (left to right): leaf trimmer; rake with spatula; root hook; coir brush; concave cutter; knob cutter; wire cutter; small, medium and large shears.

Soil and Fertilization

Akadama soil.

Bonsai soil is usually a loose, fast-draining mix of components, often a base mixture of coarse sand or gravel, fired clay pellets, or expanded shale combined with an organic component such as peat or bark. The inorganic components provide mechanical support for bonsai roots, and—in the case of fired clay materials—also serve to retain moisture. The organic components retain moisture and may release small amounts of nutrients as they decay.

In Japan, bonsai soil mixes based on volcanic clays are common. The volcanic clay has been fired at some point in time to create porous, water-retaining pellets. Varieties such as akadama, or "red ball" soil, and kanuma, a type of yellow pumice used for azaleas and other calcifuges, are used by many bonsai growers. Similar fired clay soil components are extracted or manufactured in other countries around the world, and other soil components like diatomaceous earth can fill a similar purpose in bonsai cultivation.

Opinions about fertilizers and fertilization techniques vary widely among practitioners. Some promote the use of organic fertilizers to augment an essentially inorganic soil mix, while others will use chemical fertilizers freely. Many follow the general rule of little and often, where a dilute fertilizer solution or a small amount of dry fertilizer are applied relatively frequently during the tree's growing season. The flushing effect of regular watering moves unmetabolized fertilizer out of the soil, preventing the potentially toxic build-up of fertilizer ingredients.

Pest Management

The common pests afflicting bonsai include insects both above and beneath the soil, and infections, usually fungal. A tree grown as a bonsai is subject to the pests that affect the same species full-grown, and also to pests common to other potted plants. Most pests are species-specific, so a detailed understanding of the specific bonsai species is necessary for identifying and treating most pests. The same materials and techniques used for other affected plants can be applied to the bonsai, with some relatively minor variation. Pesticide chemicals are usually diluted more for bonsai than for a larger plant, as a regular-strength application may overwhelm the smaller bonsai's biological processes.

Bonsai are sometimes marketed or promoted as house plants, but few of the traditional bonsai species can thrive or even survive inside a typical house. Most bonsai are located out of doors. The best guideline to identifying a suitable growing environment for a bonsai is its native hardiness. If the bonsai grower can closely replicate the full year's temperatures, relative humidity, and sunlight, the bonsai should do well. In practice, this means that trees from a hardiness zone closely matching the grower's location will generally be the easiest to grow, and others will require more work or will not be viable at all.

Indoors

Tropical and Mediterranean species typically require consistent temperatures close to room temperature, and with correct lighting and humidity many species can be kept indoors all year. Those from cooler climates may benefit from a winter dormancy period, but temperatures need not be dropped as far as for the temperate climate plants and a north-facing windowsill or open window may provide the right conditions for a few winter months.

References

- Gardening, science: britannica.com, Retrieved 5 February, 2019
- Mills, linn (29 january 2012). "reap benefits of container gardening". Las vegas review journal. Retrieved 1 january 2013
- Landscape, horticulture: indiaagronet.com, Retrieved 3 May, 2019
- Thompson, al (18 december 2012). "container gardening offers many benefits". Santa maria times. Retrieved 1 january 2013
- Holmgren, d. (2002). Permaculture, principles and pathways beyond sustainability. Holmgren design services, hepburn, australia. Isbn 0-646-41844-0
- Grafting as a bonsai tool Bonsaikc.com. Archived from the original on 2009-02-04. Retrieved 2009-04-28
- Prescott, david (2001). The bonsai handbook. New holland publishers ltd. P. 62. Isbn 978-1-85974-708-7

Disease Management of Horticultural Crops

6

- **Fruit and Vegetable Diseases**
- **Disease Management of Horticultural Crops**

Horticulture crops are susceptible to diverse range of diseases. The most common of them are bacterial speck, clubroot, anthracnose disease, floral malformation, leaf malformation, etc. This chapter closely examines these diseases as well as their management to provide an extensive understanding of the subject.

Fruit and Vegetable Diseases

Diseases caused by fungi and bacteria are following:

Bacterial Speck

Bacterial speck is a bacterial disease and results in small, black spots on leaves, stems and fruits of tomatoes, at all stages of growth. It is most common from winter to mid spring. Copper hydroxide may give some control.

Clubroot

Clubroot is a fungus disease that only affects Brassicas (the broccoli family). Plants are yellowish and stunted, with large malformed 'clubbed' roots. Clubroot may be severe in warm weather. Avoid growing brassicas in the same area for four years and lime the soil if it is acidic.

Leaf Spot Diseases

There are many types of leaf spot diseases that can affect beetroot, broad beans, carrots celery, peas, potatoes (early blight) silverbeet, and tomatoes (target spot). Sometimes the leaf spots cause only slight damage, but if necessary, control with mancozeb or copper hydroxide.

Damping-off

Damping-off fungal diseases such as pythium may kill small seedlings of most vegetables. Seedlings die before they emerge or soon after emergence, which results in plant collapse. Damage may occur all year, mainly in wet conditions. Do not over-water and ensure that plants are not too crowded.

Downy Mildew

In spring, with mild, humid weather, downy mildew fungus disease can cause greyish-white patches on the leaves of onions, which can droop from the tips. Cauliflowers, peas and lettuce may show white, downy growth on the underside of the leaves. With rhubarb, patches develop which tear to produce holes in the leaves. The fungus that causes downy mildew in a particular crop is specific to that crop. Control with mancozeb or copper hydroxide.

Powdery Mildew

Powdery mildew is a fungal disease. In warm, moist conditions, white patches occur on the surfaces of older leaves and leaves turn brown and shrivel. The disease is common with the cucurbit family such as cucumbers, melons, pumpkins and zucchini. It may also occur in peas and strawberries. Sulphur sprays will help to control the disease, but do not spray on hot days. Fungicides containing bicarbonate of soda also help control mildew but should be applied as soon as the disease is noticed.

Tuber Diseases

Potato tubers may be infected with superficial skin diseases, such as common scab, powdery scab and rhizoctonia. Sweet potatoes may be infected by scurf. The tubers are often edible if the infected areas are cut out. Rotate crops and plant from healthy stock.

White Mould

White mould or sclerotinia is a fungal disease that can attack most vegetables, especially beans, celery, lettuce and the broccoli family, mainly from spring to autumn. Soft brownish water rots develop mainly on the stems, followed by a fluffy white growth and small black pebble-like bodies. White mould may stay in the soil for many years and affect following crops. Do not over-crowd or over-water plants.

Wilt Diseases

Fungal diseases such as fusarium, rhizoctonia and verticillium can cause wilting and death of most vegetables, by attacking roots and basal stems. Burn diseased plant remains, rotate crops and use new stakes. Some hybrid tomato varieties have resistance to fusarium.

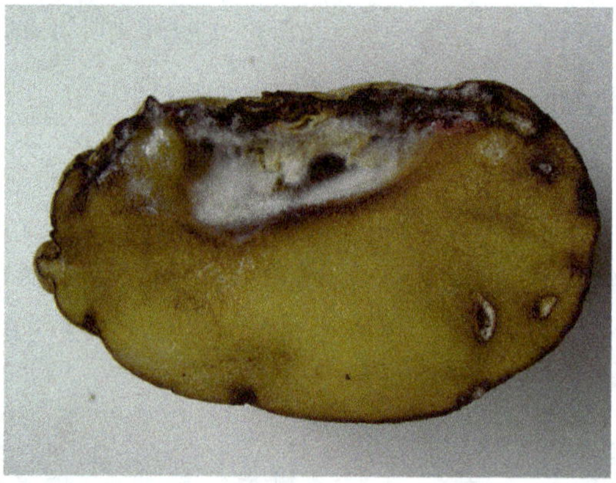

Viruses and Mycoplasmas

There are many types of viruses and mycoplasmas that can damage most vegetables. Symptoms are usually stunting, and mottled twisted leaves. There are no pesticides that can be used to kill viruses. These are often spread by aphids, thrips or leafhoppers. Reduce viruses by controlling these insects with pesticides.

Severe viruses include cucumber mosaic virus (most vegetables), zucchini yellow mosaic virus (cucumber, pumpkins, melons and zucchini), leaf roll virus and virus X and Y (potatoes), and spotted wilt virus (capsicums, lettuce and tomatoes). Big vein mycoplasma is common on lettuce in winter and does not have an insect vector. Big bud mycoplasma attacks tomatoes.

Blossom End Rot

Blossom end rot is not a disease. It occurs mainly in tomatoes, capsicums and watermelons as a dry rot at the end of the fruit. It is caused mainly by a shortage of water in hot weather and some varieties are more tolerant than others. Round varieties of watermelons are less affected by this rot.

Disease Management of Horticultural Crops

Fruit Crops

Diseases of Mango Anthracnose (Colletotrichum Gloeosporioides)

Symptoms:

- Produces leaf spots, blossom blight, wither tip, twigs blight and fruit rot.
- Small blister like spots develop on the leaves and twigs. Young leaves wither and dry Tender twigs wither and die back symptom appears.
- Affected branches ultimately dry up. Black spots appear on fruits.
- The fruit pulp becomes hard, crack and decay at ripening. Infected fruits drop.

Anthracnose disease

Leaves and inflorescence.

Fruits.

Mode of Survival and Spread:

- On dried leaves, defoliated branches mummified flowers and flower brackets.
- Contact with diseased fruit during transport and storage. The secondary spread is through airborne conidia.

Favourable Conditions:

- Temperature of 25 °C and Relative Humidity 95-97 %.

Management:

- Before storage, treat with hot water, (50-55 °C) for 15 minutes.

Mango Malformation (Fusarium Moliliforme Var. Subglutinans)

Symptoms:

- Three types of symptoms: bunchy top phase, floral malformation and vegetative malformation.

- In bunchy top phase in nursery bunch of thickened small shoots, bearing small rudimentary leaves. Shoots remain short and stunted giving a bunchy top appearance.
- In vegetative malformation, excessive vegetative branches of limited growth in seedlings. They are swollen with short internodes forming a bunches of various size and the top of the seedlings shows bunchy top appearance.
- In malformation of inflorescens, shows variation in the panicle. Malformed head dries up in black massesa and persist for long time.
- Secondary branches are transformed into number of small leaves giving a witches broome appearance.

Floral malformation.

Leaf malformation.

Mode of Spread:

- Diseased propagatives materials.

Management:

- Diseased plants should be destroyed.
- Use of disease free planting material.
- Pruning of diseased parts along the basal 15-20 cm apparently healthy portions.

Grey Blight (Pestalotia Mangiferae)

Symptoms:

- Brown spots develop on the margin and at the tip of the leaf lamina. They increase in size and become dark brown. Black dots appear on the spots which are acervali of the fungus.

Mode of Survival and Spread:

- Survive on mango leaves for over a year. Spreads through wind borne conidia.

Favourable Conditions:

- Heavy infection is noticed during the monsoon when the temperature is 20-25oC.and high humidity.

Management:

- Remove and destroy infected plant parts.

Sooty Mould (Capnodium Mangiferae)

Symptoms:

- The fungi produce mycelium which is superficial and dank.They row on sugary secretions of the plant hoppers. Black encrustation is formed which affect the photosynthetic activity.

Favourable Conditions:

- The fungus grows on the leaf surface on the sugary substances secreted by Jassids,Aphids and scale insects.

Management:

- Management should be done for insects and sooty moulds simultaneously.
- Starch dries and forms a flake which are removed along with the fungus.

Diseases of Citrus Gummosis

(Phytophthora Parasitica, P. Palmivora and P. Citrophthora)

Symptoms:

- First symptoms are dark staining of bark which progresses into the wood.
- Bark at the base is destroyed resulting in girdling and finally death of the tree.
- Bark in such parts dries, shrinks and cracks and shreds in lengthwise vertical strips.
- Later profuse exudation of gum from the bark of the trunk.
- Infection extends to crown roots.

Favourable Conditions:

- Prolonged contact of trunk with water as in flood irrigation; water logged areas and heavy soils.

Mode of Spread and Survival:

- Soil inhabitants.
- Sporangia spread by splashing rain water, irrigation water and wind.
- Irrigation water and wind.

Management:

- Injuries to crown roots or base of stem during cultural operations should be avoided.
- If lesion has girdled less than ½ the girth, remove the diseased bark with a knife along with ½" of uninvaded bark.

Canker (Xanthomonas Campestris Pv Citri)

Symptoms:

- Acid lime, lemon and grapefruit are affected. Rare on sweet oranges and mandarins.
- Affects leaf, twig and fruits. In canker, leaves are not distorted.
- Lesions are typically circular with yellow coloured; appear on both sides of leaf, severe in acid lime (difference from scab) When lesions are produced on twigs, they are girdled and die.
- On fruits, canker lesions reduce market value.

Favourable Conditions:

- Free moisture for 20 minutes, 20-30 °C.

Mode of Survival and Spread:

- Wind and rain splashes.
- Survives in infected leaves for 6 months.
- Injury caused by leaf miner helps the entry of the bacterium.

Management:

- Control leaf miner when young flush is produced.
- Prune badly infected twigs before the onset of monsoon.

Tristeza or Quick Decline (Virus)

Symptoms:

- Lime is susceptible both as seedling or buddling on any root stock.
- But mandarin and sweet orange seedlings or on rough lemon, trifoliate orange, citrange; Rangpur lime root stocks tolerant; susceptible root stocks are grapefruit and sour orange.
- In sweet orange or mandarin on susceptible root stocks, leaves develop deficiency symptoms and absise.
- Roots decay, twigs die back. Fruit set diminishes; only skeleton remains.
- Fine pitting of inner face of bark of sour orange stock.
- Grapefruit and acid lime are susceptible irrespective of root stock.
- Tree stunted and dies yield very much reduced. Fruits are small in size.

Mode of Spread:

- Use of infected bud wood *Toxoptera citricida* (aphid) is the important vector.

Management:

- For sweet orange and mandarin. Avoid susceptible root stocks.
- For acid lime, use seedling preimmunised with mild strain of tristeza.

Greening (Liberobactor Asiaticum) (Phloem Limited Bactria-like-organism)

Symptoms:

- This disease affects almost all citrus varieties irrespective of root stock.
- Stunting of leaf, sparse foliation, twig die back, poor crop of predominantly greened, worthless fruits.
- Sometimes only a portion of tree is affected. A diversity of foliar chlorosis.
- A type of mottling resembling zinc deficiency often predominates.
- Young leaves appear normal but soon assume on outright position, become leathery and develop prominant veins and dull olive green colour. Green circular dots on leaves.
- Many twigs become upright and produce smaller leaves.
- Fruits small, lopsided with curved columella. The side exposed to direct sunlight develops full orange colour but the other side remain dull olive green.
- Low in juice and soluble solids, high in acid. Worthless either as fresh fruit or for processing.

Seeds poorly developed, dark coloured, aborted.

Mode of Spread:

- Infected budwood; psyllid vector-*Diaphorina citri.*

Management:

- Control psyllids with insecticides.
- Use pathogen free bud wood for propagation.

Diseases of Guava

Anthracnose (Colletotrichum Gloeosporioides)

Symptoms:

- Symptoms of this disease are observed on mature fruits on the tree.
- The characteristic symptoms consist of sunken, dark colored, necrotic lesions. Under humid conditions, the necrotic lesions become covered with pinkish spore masses.
- As the disease progresses, the small sunken lesions coalesce to form large necrotic patches affecting the flesh of the fruit.

Management:

- Control measures are needed in commercial guava production.
- The use of resistant cultivars provides the most efficient tactic in disease management.

Guava Rust (Puccinia Psidii)

Symptoms:

- The pathogen can affect foliage, young shoots, inflorescences and fruit of guava.
- Typical symptoms associated with this disease include distortion, defoliation, reduced growth and if severe, mortality.
- On fully expanded leaves, dark bordered, roughly circular brown lesions with yellow halos develop.

Management:

- Control of guava rust is based on the use of fungicides.
- Scouting fields for onset of disease or during the times of year when environmental conditions are favorable for pathogen infection are recommended so that proper and timely fungicide applications can be made.
- In addition, proper cultural tactics such as proper fertilization, irrigation, pruning and

sanitation aide in achieving a healthy, vigorously growing tree less vulnerable to disease pressures.

Guava Diseases caused by Nonfungal Agents (Cephaleuros Virescens) Algal Leaf Spot

Symptoms:

- Disease symptoms are exhibited on both abaxial and adaxial leaf surfaces as orange, rust-colored, dense silky tufts ranging from 5 to 8 mm in diameter.
- Upon scraping away these spots, a thin, grayish white to dark-colored, necrotic crust remains on the leaf.
- These spots usually come together to form large irregular patches on a leaf. As the spots mature they take on a dull, grayish green color.
- Twigs and branches are also affected causing the bark to crack due to the growth and expansion of the pathogens filaments into the cortical tissues of the host.

Management:

- Algal leaf spot can be reduced by maintaining tree vigor with cultural techniques such as proper fertilization and irrigation, proper pruning to enhance air circulation within the canopy and sunlight penetration, managing weeds and wider tree spacing. Managing insect, mite and other foliar diseases increases tree vigor and lessens susceptibility to this disease.

Diseases of Apple Scab (Venturia Inaequalis)

Symptoms:

- Symptom appears on leaves and fruits.
- On lower side of the leaf lesion appear as olivaceous spots which turn dark brown to black and become velvety.
- On young foliage, the spots have a radiating appearance with a feathery edge.
- On older leaves the lesions are more definite in outline.
- The lesion may form a convex surface with corresponding concave area on the opposite side.
- In severe infection leaf blade curved, dwarfed and distorted.
- Fruits show small, rough, black circular lesions.
- The centre of the spots become corky and on mature fruits, yellow halo is seen around the lesions.

Mode of Spread and Survival:

- Primary spread through ascospores and secondary spread by air borne conidia.
- Favourable conditim.
- Cool and moist conditions due to rain or show at higher elevation and shady portion.

Management:

- Clean cultivation, collection and destruction of fallen leaves.

Fire Blight (Erwinia Amylovora)

- The initial symptom usually occurs on flowers, which become water soaked, then shrivel turn brownish to black in colour and fall or remain hanging in tree.
- The symptom spread to leaves twigs. Terminal twigs wilt from hip to downward and also spread to branches fruits becomes water soaked, turns brown, shrivels and findly becomes black.
- Oozing may be seen in the affected area.

Survival and Spread:

- The bacterium survives in infected branches and twigs. Secondary spread by insects and rain splash.

Favourable Condition:

- Temperature above 24 °C and heavy rain.

Management:

- Removal and destruction of affected parts.
- cutting of blighted twigs and root sprouts in summer.

Soft Rot (Penicillium Expansum)

Symptom:

- Young spots starts from stem end of the fruit as light brown watery rot. As the fruit ripens area of the rotting increases.

- Skin becomes wrinkled.
- A peculiar musty odour is emitted.
- Under humid condition a bluish green sprorulating growth appears.

Mode of Spread and Survival:

- Infection take place by wounds in the skin caused by insects and during handing in storage and transport.

Management:

- Careful handling of fruits without causing any wounds.

Vegetable Crops

Diseases of Tomato Damping Off (Pythium Aphanidermatum)

Symptom:

- This is one of the worst diseases of tomato occurring in the nursery.
- Damping off of tomato occurs in two stages, i.e. the pre-emergence and the post-emergence phase.
- In the pre-emergence the phase the seedlings are killed just before they reach the soil surface.
- The young radical and the plumule are killed and there is complete rotting of the seedlings.
- The post-emergence phase is characterized by the infection of the young, juvenile tissues of the collar at the ground level.
- The infected tissues become soft and water soaked. The seedlings topple over or collapse.

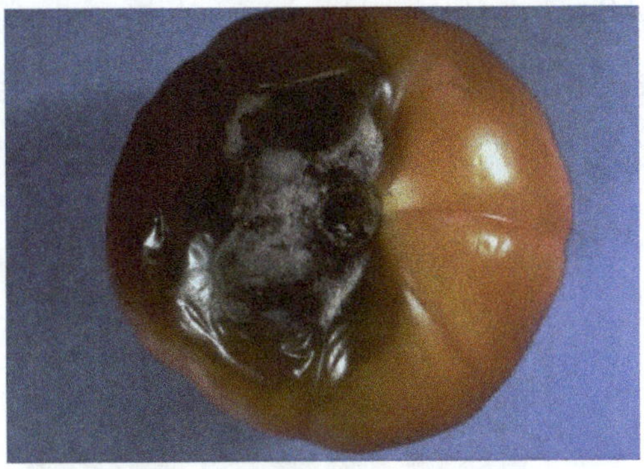

Control:

- Seed treatment with fungal culture *Trichoderma viride* (4 g/kg of seed).

Early Blight (Alternaria Solani)

Symptom:

- This is a common disease of tomato occurring on the foliage at any stage of the growth.
- The fungus attacks the foliage causing characteristic leaf spots and blight. Early blight is first observed on the plants as small, black lesions mostly on the older foliage.
- Spots enlarge, and by the time they are one-fourth inch in diameter or larger, concentric rings in a bull's eye pattern can be seen in the center of the diseased area.
- Tissue surrounding the spots may turn yellow. If high temperature and humidity occur at this time, much of the foliage is killed.
- Lesions on the stems are similar to those on leaves, sometimes girdling the plant if they occur near the soil line.
- Transplants showing infection by the late blight fungus often die when set in the field. The fungus also infects the fruit, generally through the calyx or stem attachment.

Control:

- Removal and destruction of the affected plant parts. Practicing crop rotation helps to minimize the disease incidence.

Fusarium Wilt (Fusarium Oxysporum f. sp. Lycopersici)

Symptom:

- This is one of the worst diseases of tomato occurring mostly in the nurseries. The first symptoms of the disease are clearing of the veinlets and chlorosis of the leaves.
- The younger leaves may die in succession and the entire may wilt and die in a course of few days. Soon the petiole and the leaves droop and wilt.
- In young plants, symptom consists of clearing of veinlet and dropping of petioles. In field, yellowing of the lower leaves first and affected leaflets wilt and die.
- The symptoms continue in subsequent leaves. At later stage, browning of vascular system occurs. Plants become stunted and die.

Control:

- The nursery should be regularly inspected for wilt infected plants.
- The affected plants should be removed and destroyed.
- Prior to planting the beds should be drenched with Carbendazim (0.1%) and the seeds should be treated with the Thiram (2.5 kg/ha).
- Crop rotation with a non-host crop such as cereals helps to reduce the disease inoculum.

Septoria Leaf Spot (Septoria Lycopersici)

Symptom:

- The plant may be attacked at any stage of its growth. The disease is characterized by numerous, small, grey, circular leaf spots having dark border.

Control:

- Removal and destruction of the affected plant parts.

Bacterial Wilt (Pseudomonas Solanacearum)

Symptom:

- This is one of the most serious diseases of tomato crop. Relatively high soil moisture and soil temperature favour disease development.
- Characteristic symptoms of bacterial wilt are the rapid and complete wilting of normal grown up plants.
- Lower leaves may drop before wilting. Pathogen is mostly confined to vascular region; in advantage cases, it may invade the cortex and pith and cause yellowbrown discolouration of tissues.
- Infected plant parts when cut and immersed in clear water, a white streak of bacterial ooze is seen coming out from cut ends.

Control:

- Crop rotations, viz., cowpea-maize-cabbage, okra-cowpea-maize, maize- cowpea-maize and finger millet-egg plant are reported effective in reducing bacterial wilt of tomato.
- Seedling treatment with Streptocycline (1 g/40 litres of water) for 30 min protects the seedlings in the initial stages of growth.

Bacterial Leaf Spot (Xanthomonas Campestris pv. Vesicatoria)

Symptom:

- Moist weather and splattering rains are conducive to disease development. Most outbreaks of the disease can be traced back to heavy rainstorms that occur in the area.

- Infected leaves show small, brown, water soaked, circular spots surrounded with yellowish halo.
- On older plants the leaflet infection is mostly on older leaves and may cause serious defoliation.
- The most striking symptoms are on the green fruit. Small, water-soaked spots first appear which later become raised and enlarge until they are one-eighth to one-fourth inch in diameter.
- Centers of these lesions become irregular, light brown and slightly sunken with a rough, scabby surface.
- Ripe fruits are not susceptible to the disease. Surface of the seed becomes contaminated with the bacteria, remaining on the seed surface for some time.
- The organism survives in alternatehosts, on volunteer tomato plants and on infected plant debris.

Control:

- Bacterial spot is difficult to control once it appears inthe field.
- Disease-free seed and seedlings should always be used and the crop should be rotated with non-host crops so as to avoid last years crop residue.
- Spraying with Agrimycin-100 (100 ppm) thrice at 10 days intervals effectively controls the disease.

Tomato Mosaic Virus (TMV)

Symptom:

- The disease is characterized by light and day green mottling on the leaves often accompanied by wilting of young leaves in sunny days when plants first become infected.
- The leaflets of affected leaves are usually distorted, puckered and smaller than normal. Sometimes the leaflets become indented resulting in "fern leaf" symptoms.
- The affected plant appears stunted, pale green and spindly.
- The virus is spread by contact with clothes, hand of working labour, touching of infected plants with healthy ones, plant debris and implements.

Control:

- Seeds from disease free healthy plants should be selected for sowing. The seeds should be thoroughly rinsed and dried in shade.
- In the nursery all the infected plants should be removed carefully and destroyed. Seedlings with infected with the viral disease should not be used for transplanting.
- Crop rotation with crops other than tobacco, potato, chilli, capsicum, brinjal, etc. should be undertaken.

Tomato Leaf Curl Virus (TLCV)

Symptom:

- This disease is transmitted by whitefly (Bemisia tabaci). It is one of the most devastating diseases of tomato.
- Leaf curl disease is characterized by severe stunting of the plants with downward rolling and crinkling of the leaves. The newly emerging leaves exhibit slight yellow colouration and later they also show curling symptoms.
- Older leaves become leathery and brittle. The nodes and internodes are significantly reduced in size.
- The infected plants look pale and produce more lateral branches giving a bushy appearance. The infected plants remain stunted.

Control:

- The affected plants should be removed and destroyed. Alternate or collateral hosts harboring the virus causing this disease is removed at the time of weeding or earthing up operations to minimize the spread of the disease.
- Checking the white fly population can reduce the disease incidence.

Tomato Spotted Wilt Virus (TSWV)

Symptom:

- The spotted wilt virus is transmitted through thrips (*Thrips tabaci, Frankliniella schultzi* and *F. occidentalis*).
- This disease is similar to streak in that it causes streaking of the leaves, stems and fruits. Numerous small, dark, circular spots appear on younger leaves.
- Leaves may have a bronzed appearance and later turn dark brown and wither.
- Fruits show numerous spots about one-half inch in diameter withconcentric, circular markings. On ripe fruit, these markings are alternate bands of red and yellow.

Control:

- The affected plants should be removed and destroyed.
- Alternate or collateral hosts harboring the virus causing this disease is removed at the time of weeding or earthing up operations to minimize the spread of the disease.

Diseases of Brinjal

Bacterial Wilt (Pseudomonas Solanacearum)

Symptoms:

- Bacterial wilt symptoms on leaf surface Wilting, stunting, yellowing of the foliage and finally collapse of the entire plant are the characteristic symptoms of the disease.
- Lower leaves may droop first before wilting occurs.
- The vascular system becomes brown.
- Bacterial ooze comes out from the affected parts.
- Plant show wilting symptoms at noontime will recover at nights, but die soon.

Control:

- Pant samrat variety is tolerant.
- Crop rotation with cruciferous vegetables such as cauliflower help in reducing the disease incidence.
- Fields should be kept clean and effected parts are to be collected and burnt.
- The disease is more prevalent in the presence of root knot Nematodes, so control of these nematodes will suppress the disease spread.

Cercospora Leaf Spot (Cercospora Solani-Melongenae and C. Solani)

Symptoms Leaf Spot

Symptoms:

- The leaf spots are characterized by chlorotic lesions, angular to irregular in shape, later turn grayish-brown with profuse sporulation at the centre of the spot.
- Severely infected leaves drop off prematurely, resulting in reduced fruit yield.

Control:

- Pant Samrat variety is resistant to both the leaf spots.
- Diseases can be managed by growing resistant varieties.

Tobacco Mosaic Virus (TMV)

Symptoms:

- Mosaic mottling of leaves and stunting of plants are the characteristic symptoms of potato virus Y Mosaic symptoms are mild in early stages but later become severe.
- Infected leaves are deformed, small and leathery. Very few fruits are produced on infected plants.
- The important symptom produced by tobacco mosaic virus is conspicuous mottling of leaves.
- Leaves also develop blisters in advanced cases. Severely infected leaves become small and misshapen. Plants infected early remain stunted.
- PVY is easily sap transmitted.
- It is transmitted in the field through aphids, *Aphis gossypii* and *Myzus persicae* and perpetuates on weed hosts like *Solanum nigrum* and *S.xanthocarpum.*
- TMV is transmitted by sap, contaminated implements and clothes, soil debris and hands of labour.
- It can perpetuate on many cultivated plants like cucurbits, legumes, pepper, tobacco, tomato and weed hosts. The virus survives in plant debris in soil.

Control:

- Destroy all weeds and avoid planting cucumber, pepper, tobacco, tomato near brinjal seed beds and field.
- Wash hands with soap and water before working in seed beds.
- Prohibit smoking or chewing of tobacco who are handling brinjal seedlings.

Collar Rot (Sclerotium Rolfsii)

- The disease occasionally occurs in serious form.

Symptoms:

- The lower portion of the stem is affected from the soil borne inoculum (sclerotia).
- Decortication is the main symptom.
- Exposure and necrosis of underlying tissues may lead to collapse of the plant.
- Near the ground surface on the stem may be seen the mycelia and sclerotia.
- Lack of plant vigour, accumulation of water around the stem, and mechanical injuries help in development of this disease.

Control:

- Seed treatment with 4 g of *Trichoderma viride* formulation per kg seed will help in reducing the disease.
- Collection and destruction of diseased parts and portions of the plant.

Diseases of Cucurbits

Fusarium wilt of water melon (Fusarium oxysporum f. sp. Niveum), Muskmelon – F. o. fsp. Melonis.

Symptoms:

- The disease attack the plant at all stages of plant growth.
- When young seedlings are infected they damp off and die.
- In older plants the plant wilt and die within 10 days.
- Vascular discolaration can be seen. In wet weather, pinkish or whitish fungal growth can be seen on dead stems.

Mangement:

- Collection and burning of infected plant.
- Seed treatment and soil application with *T. Viride or P. fluorescens.*
- Use of disease free seeds.

Root Rot of Muskmelon (Pythium Aphendermatum)

Root rot of pumpkin and squash (*Fusarium solani f.sp. cucurbitae).*

Root Rot of Watermelon, Cucumber (Pythium Irregulare and P.ultimum).

Symptoms:

- The roots of the affected plants appear water soaked with sunken darkened lesions. The crown of the plants is girdled and entire plant is collapsed. Fruits become rotted by the fungus.

Management:

- As like wilt the management practice should be followed.

Verticillium Wilt (Verticillium Albo-atrum)

Symptoms:

- The leaves of the plant become yellow from base of the plant upwards and the whole plant wilts. If the stem is cut open, brown discoloration can be seen.

Anthracnose (Collectorichum Lagenarium)

Symptoms:

- Small yellowish or water soaked areas appears on leaves which enlarge rapidly and turn brown in most cucurbits but black in water melon. The affected bart break and fall of or the whole leaf dries. When the infection occurs in stem, the whole vine is killed. When infection occurs in fruit pedicle, the young fruit may be darken, shrivel and dry. Circular black cankers appear on fruits. The spots may be up to 5 cm dia based on the host and environment. In the sunken black spots salmon coloured spores can be seen.

Management:

- Collection and destruction of infected plant.

Alternaria Leaf Spot/Target Leaf Spot/Macrosporium Blight (Alternaria Cucumerina and Macrosporium Cucumerinum)

Symptoms:

- Small, circular spots appear on leaves. As the spot enlarges concentric rings can be seen on infected area. The fungus also causes fruit rot.

Management:

- Use of disease free seeds.

Cercospora Leaf Spot (Cercospora Citrullina)

Symptoms:

- Small block circular spots with grey centre appear on leaves. Severely infected leaves fall off. The fruit size is reduced.

Management:

- Collection and burning of infected leaf.

- Crop rotation.

Downy Mildew (Pseudoperonospora Cubensis)

Symptoms:

- Symptoms resembling mosaic viz, pale green areas separated by dark green areas appear on upper surface of leaf. During wet season, corresponding lower surface is covered with faint purplish fungal growth. The entire leaf dries up quickly.

Management:

- Use of bed system with wide spacing with good drainage and air movement and exposure to sun help to check the disease development.

Angular Leaf Spot (Pseudomonas Syringae Pv. Lachrymans)

Symptoms:

- Water soaked lesions appear on leaves which are confined to veins. The turn grey to tan and form an exudates in the lower surface. The spots loosen and fall out. Infected fruits develop a brown, circular, superficial, firm rot which causes rapid detorition. The rot may extend into flesh.

Management:

- Use of disease free seed and crop rotation.

Fruit Rot (Pythium Aphenidermatum)

Symptoms:

- Initially the skin of the fruit shows soft, dark green water soaked lesions which gradually develop into a watery soft rot. Cottony mycelium develop on the affected portions. The fruit in contract with the soil is attacked easily.

Management:

- Keep fruits not in touch with soil.

Other Fruit Rots of Watermelon

- Apergillus fruit rot – *Aspergillus flavus.*
- Curvularia fruit rot – *Curvularia ovoidae.*
- Myrothecium fruit rot – *Myrothecium roridum.*
- Rhizopus fruit rot – *Rhizopus oryzae.*
- Diplodia fruit rot – *Diplodia natalensis (*also affects cucumber).

Viral Disease

- Cucumber mosaic virus – Vector- aphids (*A. craccivora M.persicae).*

Symptoms:

- Mosaic mottling, leaf distortion, stunting of the plants, shortened internode and petoles. Infected plants bear only very few flowers. In fruits mottled yellowish green areas can be seen.

Management:

- Removal and burning of infected plants.
- Eradication weeds hosts.

Other Viral Diseases

- Watermelon mosaic virus – watermelon and muskmelon.
- Squash mosaic – Squash and pumpkin.
- Mosaic disease – Bottle gourd, snake gourd, squash, ribbed gourd.

Management:

- Removal and burning of infected plants.
- Eradication weeds hosts.

Phyllody – Phytoplasma (Vector – Leaf Hopper)

Symptom:

- It affects bitter gourd, bottle gourd, cucumber, ridge gourd and snake gourd. The symptoms are shortening of internodes and phyllody of normal flowers. Corolla, androecium and gynecium are transferred into green leaf like structures.

Management:

- Removal of infected plants.
- Clean cultivation without weed host.

Diseases of French Bean

Anthracnose

Symptoms:

- Brown canker on pods.

Management:

- Use disease free seeds.

Angular Leaf Spot

Symptoms: On the leaves there are angular red brown spots.

Management:

- Use healthy seeds.
- Adopt 2 year crop rotation in infected field.

Common Mosaic

Symptoms:

- Leaf turned curved due to etiolation and dry down near veins. There is less number of pods and seeds.

Management:

- Grow resistant varieties e.g. contender.

Diseases of Carrot

Black Root Rot (Thielaviopsis Basicola)

- This can be a problem disease on carrots grown on muck or high organic soils.
- Washed carrots may show black scattered lesions when stored in polythene bags.

Management:

- Root damage and high storage temperatures appear to favor disease development.

Cavity Spot (Pythium spp.)

Symptom:

- Cavity spot occurs on both parsnips and carrots.
- Tonnage of the crop is not reduced but the cavities make the roots unmarketable.
- The disease occurs on carrots grown on mineral or peat (muck) soils.

Management:

- No control for this disease is known other than growing carrots on new land free of this problem disease.

Common Scab (Streptomyces Scabies)

Symptom:

- This is the same scab that attacks potatoes and other root crops.
- Alkaline soils and dryish high organic soils are conducive to the development of the superficial disease.
- Scab lesions on the carrot surface.

Management:

- Avoid potato fields and lower the soil pH if above 7 or more.

Cercospora Leaf Spot (Cercospora Carotae)

Symptom:

- This disease can be as destructive as Alternaria leaf blight.
- Again the disease is both seed- and residue-borne.
- Younger leaves are more susceptible than older leaves.

Management:

- Ploughing under crop residues, crop rotation and foliar fungicides are all control recommendations.

Sclerotinia Rot - White Mold (Sclerotinia sclerotiorum)

Symptom:

- The infectious ascospores of this fungus are abundant as a consequence of the build-up of the disease on millions of acres of canola and bean crops.
- Carrots may show little or no damage incidence in the field but following washing and storage white mold outbreaks often occur on the stored roots.
- Only a small percentage of the roots may be initially infected but the fungus mycelium can move very rapidly from carrot to carrot.
- In a matter of weeks the whole storage container may become a mass of white mold and black sclerotia surrounding each and every carrot.

Management:

- Frequent inspection in storage, low temperatures, aeration.

Bacterial Leaf Blight (Xanthomonos Campistris pv. Carotae)

- A seed-borne bacterial disease that causes leaf spotting similar to Cerospora or Alternaria blights.

Diseases of Crucifers

Cabbage, cauliflower, turnip, broccoli are some of the important cruciferous crops.

Club Root (Plasmodiophora Brassicae)

- The above ground symptoms are, yellowing of leaves, stunting and later the diseased plant die.
- Young plants die with in a short time while older plants fail to produce marketable heads.
- On roots and root lets, spindle shaped with thick centre and tapered ends giving a club shaped malformation is developed.
- The club shaped structered cells are invaded by the secondary,weakly parasitic organisms and the cells are disintegrated which produce toxins and the plants are killed.

Management:

- Avoiding infected field.

Black Rot (Xanthomonas Campestris Pv. Campestris)

- On leaves 'V' shaped chlorotic to yellow lesions develop from the leaf margins.
- They become black later. The veins and veinlets become black.Systemic infection spreads to the root.
- The vascular bundle in the stem becomes black followed by integration of fleshy stalk.
- The attack of soft rot organisms cause further detoriation.
- The cabbage heads and cauliflower curds are invaded and become discolored.

Management:

- Use of healthy seed.
- Spraying with Agrimycin 200 ppm.

Alternaria Leaf Spot/Black Spot/Brown Spot (Alternaria Brassicicola)

Symptom:

- Small dark coloured lesions up to 1.0 cm in dia appear on leaves.
- In humid condition conidiophores appear on the lesion in concentric rings.
- Linear spots also appear on petioles, stems and seed pods.
- On cauliflower curds brown discoloration occurs.
- In case of *A.brassicae* the spots are smaller in dia and lighter in colour.

Favourable condition:

- Hot moist weather, temperature 26° C. Dew or rain for 9 hrs is essential for infection.

Management:

- Hot water treatment of seeds at 500C for 30 minutes.
- Crop rotation, avoiding overhead irrigation.

Wire Stem

- Wire stem gets its name from symptoms that occur on the stem at the soil level. A dark, watersoaked lesion initially appears on the stem.
- Later stems become wiry and slender at the point of the lesion.
- Diseased crucifer plants transplanted to the field grow poorly, are stunted, and may eventually die, especially if there is inadequate moisture shortly after transplanting.

- If infected plants remain alive, the stem becomes tough and woody. Plants that survive usually mature late and fail to produce a marketable head.

Bottom Rot

- Bottom rot is a disease of mature cabbage.
- After cabbage transplants become large enough to begin to shade the ground, the disease appears on the lower side of the head leaves that are in contact with the soil.
- The midrib is often the first part of the leaf attacked.
- Resulting lesions are sunken, black, and sharply elliptical with the long side of the lesion parallel to the side of the midrib.
- Lesions may dry out and become papery brown in appearance if the weather becomes dry.
- The surface of the lesion may be covered with a sparse, weblike mycelium.
- Eventually a general black decay occurs at the base of the leaf.
- The tip of the leaf then turns yellow and then the entire leaf wilts.
- Infected leaves are shed and only a naked stalk, capped with a small head, remains.
- Bottom rot becomes a head rot if temperatures are warm and the relative humidity is high.
- A weblike mycelium develops between diseased leaves.
- Infected leaves are eventually covered with small, brown, fungal structures called sclerotia, which can persist on plant debris in the soil for long periods of time.
- The stem remains unaffected, thus, the head remains upright.

Cultural Control

- Choose a fertile, well-drained site for the plant bed. Poor drainage favors seedling disease.
- Avoid excessive amounts of nitrogen fertilizer. Succulent plants are more susceptible to infection.
- Seed crucifers when the soil temperature reaches 69 °F (21 °C) and seed as shallowly as possible so that germination and emergence are rapid.
- Discard transplant seedlings that show symptoms of wire stem.
- Avoid banking or throwing soil up around plants during field cultivation.
- Harvest cabbage heads early. Cabbage leaves become more susceptible the longer they stand in the field.

Survival:

- Sources of the fungus include infested seeds, cruciferous weeds, and residues of cruciferous crops remaining in or on the soil.

- Survival in crop residues in soil for 3 years.

Management:

- Use of disease free seed.
- Rotation with non cruciferous crops for 4 years where the disease is severe.
- Providing adequate drainage facility.
- Soil fumigation with chloropicrin etc.

Cabbage Yellows/Fusarium Wilt of Cabbage (F. Oxysporum Conglutians)

Symptom:

- The plant become yellow in colour.
- The wilting of leaves may be more prominent on one side of plant that the other.
- Leaves dries up. When the infected stem is cut open brown discoloration will be seen.

Management:

- Use of disease free seed.Crop rotation, field sanitation.
- Use of resistant varieties.

Ring spot of Cabbage (Mycosphaerella Brassicicola)

Symptom:

- Outer leaves are severely affected.
- Brown to tan spots of 1-2 cm dia. appear on leaves.
- Central portions are grey in colour with numerous fruiting bodies appear in concenric rings.
- The spots are surrounded by a green band which remains great even after the whole leaves turn yellow.
- Diseased leaves fall off. When the spot affects head, marked value is reduced.

Management:

- Collection and destruction of plant debris.
- Hot water treatment of seeds 45 °C for 20 min.

Stalk Rot (Sclerotinia Sclerotiarum)

Symptom:

- In cabbage water soaked spots appear on stem and leaves near the ground level.

- The leaves wilt and plant collapses with in 10-15 days. When head is infected cottony white fungal growth is seen with numerous hard black sclerotia.
- In cauliflower yellowing starts from tip of the leaves to down wards till the entire leaves are covered.
- The leaves shed prematurely.
- The rotting progress to stem and the stem girdles and the stem rots upto the curd region.
- Curd are also affected. White fluffy mycelial growth with numerous sclerotia in the affected portions can be seen..

Management:

- Crop rotation with paddy.

Disease of Onion

Basal Rot (Fusarium Oxysporum f. Cepae)

- This disease is worldwide and again caused by a specialized form of *Fusarium.*
- This disease can begin in the field and continue on in storage.

Pink Root (Pyrenochaeta Terrestris and Fusarium spp.)

Management:

- Infection is usually spread by onion sets. No other control procedures are known.

Rot (Penicillium spp.)

Management:

- Several fungal and bacterial diseases are capable of causing storage rots in onions, particularly if the crop is inadequately or poorly cured following harvest.
- Adequate ventilation must also be supplied during storage.

Disease of Garlic

Macrophomina Rot (Macrophomina Phasoolina)

Garlic types 1069-9,1069-5,1069-6,1075-6,were resistant and 1029,1037,1040, 1073 -3, 073-5, 1073-7, 1073-9, 1073-10, 1073-14 and 1073-17 were moderately resistant.

Pink Root (Pyrenochaeta Tattestris)

Pink root of garlics is caused by the fungus. (Hansen) Gorenz, walker and Larson *Pehoma terrestris* (Hanson) it has been reported from all the garlic growing areas.

Symptoms:

- The fungus attacks onion from the seedling stage onwards.
- The roots are affected and they turn pink or reddish and sometimes darken to a red or purple colour black spores form on the diseased roots which eventually shrivel and die.
- Diseased plants can be easily pulled. The above ground symptoms are shunting and yellowing tip burn and die back of the leaves.
- Affected seedlings may be killed. Older plants are not normally killed but bulb formation is affected and yields are low.
- Bulbs are not attacked although the outer scales may be penetrated.
- New roots are formed throughout the season and these may be infected and killed successively.

Spread of the Disease

The pathogen is soil borne and infection is mostly from mycelia in the soil. Optimum soil temperatures for disease development are 24-28 °C.

Management:

- Long rotations should be practiced to prevent build up of inoculum in the soil.

Diseases of Cassava

Indian Cassava Mosaic Virus Disease (ICMV)

Symptoms:

- Mosaic symptom, reduction in leaf size, misshapened twisted leaves.
- The affected plants stunted, tuber splitting and yield reduction.

Management:

- Selection of setts from healthy plants.
- Use of resistant varieties like MnGa-1.

Sett Rot/Stem Rot – Diplodia Natalensis (Botryodiplodia Theobromae)

- It is found in stored setts and new planting in dry areas.
- Vascular strands show black discoloration and necrosis radiating from wounds.
- Epidermel blisters are produced under which the infected dark brown or black tissues can be seen.
- Masses of black pycnidia can be seen when the blister rupture.

Management:

- Selection of healthy setts.

Sweet Potato

Black Rot (*Ceratocystis Fimbriata)*

Symptoms:

- Small, circular, slightly sunken, dark brown spots are the initial symptoms of black rot.
- Spots enlarge and appear greenish black to black when wet and grayish black when dry. Within the spots are small, black fungal structures (perithecia) with long necks which appear to the naked eye as dark bristles. The rot usually remains firm and shallow.
- If secondary fungi or bacteria invade the tissue however, the flesh beneath the spot turns black, and this blackened area may extend to the center of the root.
- Tissue near the discolored area may have a bitter taste. Eventually, the entire root may rot. Roots may appear healthy at harvest but rot in storage, during transit, or in the market.

Management:

- Control black rot with crop rotation, since most crops are unaffected by the disease.
- Disinfect seedbeds if a clean site is unavailable.
- Propagate plants from healthy stem cuttings.
- Cure roots immediately after harvest. (Cure roots at 85 to 95 °F and 85 to 90 percent relative humidity for 5 to 10 days).
- Do not wash and package roots showing symptoms of black rot.
- Decontaminate equipment that comes into contact with an infected crop.

Rhizopus Soft Rot (Rhizopus Stolonifer)

Symptom:

- Infection and decay commonly occur at one or both ends of the root, although infection occasionally begins elsewhere.
- Rotting may be inhibited under dry conditions, but under humid conditions the affected sweet potatoes become soft and watery, and the entire root rots within a few days.
- If the humidity is high, the sweet potatoes become heavily "whiskered" with a grayish black fungal growth. This feature distinguishes Rhizopus soft rot from other storage rots.
- The color of the root is not significantly altered, but an odor is produced that attracts fruit flies to the area.

- Infection is especially likely if the relative humidity is between 75 and 85 percent during storage or transport. Also, the longer roots are stored, the more susceptible they become.
- Chilling and heat damage also predispose sweet potatoes to infection. Soft rot is very destructive when sweet potatoes are washed, packed, or shipped to market during cold weather.

Management:

- Carefully handle sweet potatoes during harvest to prevent unnecessary wounding. This is the most important control method for soft rot.
- Properly cure roots immediately after harvest.
- Store roots at 55 to 60 degrees F.
- Avoid handling stored roots because handling can create new wounds. Recuring is one possible solution to this problem.
- Apply a recommended fungicide after harvest.
- Do not allow sweet potatoes to be exposed to sunlight for extended periods (to prevent heat damage) or to be chilled in the field.

Bacterial Soft Rot (Erwinia Chrysanthemi)

Symptoms:

- Roots are affected in the field, or more commonly in storage, by a soft rot that turns diseased tissue light brown and watery.
- Lesions on storage roots often have a dark brown margin. Some storage roots appear healthy from the outside but are decayed internally.
- Infected roots show black streaks in the vascular tissue and eventually undergo a soft, moist decay.
- Mother roots often decay in plant beds. In the field, brown to black, water-soaked lesions appear on stems and petioles. Eventually, the stem may become watery and collapse, causing the ends of vines to wilt.
- Usually, one or two vines may collapse, but occasionally the entire plant dies.

Management:

- Carefully handle sweet potatoes during all stages of production. This is the most important control method for bacterial soft rot.
- Select mother roots from fields free of the disease.
- Cull roots infected during storage.

- Use vines cut above the soil surface for transplanting.
- Use a handling system that does not involve immersion of sweet potatoes in water.

Charcoal Rot (Macrophomina Phaseoli)

Symptoms:

- In the field, brown to black, water-soaked lesions appear on stems and petioles. Eventually, the stem may become watery and collapse, causing the ends of vines to wilt.
- Usually, one or two vines may collapse, but occasionally the entire plant dies.
- Charcoal rot, caused by the fungus, can cause losses of sweet potatoes in storage, but serious losses seldom occur. The disease is sometimes confused with black rot and Java black rot.
- Symptoms in storage begin as a reddish brown to brown, firm, moist rot, initially restricted to the area just beneath the sweet potato skin.
- As the decay progresses, the pathogen moves toward the center of the sweet potato, causing further rot. Two distinct zones become apparent within the infected tissue.
- The leading edge continues as a reddish brown decay, and a zone of black develops behind the zone of active decay.
- Although the lesions are sometimes restricted, charcoal rot usually consumes the entire root, which eventually dries, becoming hard and mummified.

Management:

- Properly cure sweet potatoes immediately after harvest to reduce the incidence of charcoal rot.

Spice Crops

Diseases of Black Pepper

Diseases of Pepper: Phytophthora foot rot / quick wilt (Phytopthora capsici).

Leaf Infection

Water soaked lesions with fimbriate margin develop from margins of leaves or in the centre of the leaf. The defoliation occurs in severe infection.

Die Back

The aerial branches get infected at any point. At the site of infection of branch, the discoloration occurs and rotting progress upwards and downwards resulting in die-back symptoms. The lateral branches of the affected vines break off at the nodes and fall off.

Foot Rot or Collar Rot

The stem near the ground level get infected and the rotting and death of vine occurs with in 2-3 weeks. The affected portion emits bad odour. The necrosis progress down wards to the underground stem and to the root system.

Root Rot

The infection starts at main root or at feeder root. The leaves become yellow and defoliate.

Management:

- Selection of healthy nursery material.
- Provide good drainage.
- The tender runner shoots and leaves in the ground should be pruned off to avoid direct contract to the soil.
- Application of neem cake and *Trichoderma viride or P. fluorescens.*

Diseases of Cardamom

Damping off / rhizome rot / clamp rot (*Pythium vexans, Fusarium oxyoporum, Rhizoctonia solani*).

Symptoms:

Infected leaves become pale, yellow and ultimately the young leaves die. Older leaves die prematurely and new shoots that arise are weak, decay and the rhizomes rot at the base of the stem. The diseased shoot can be pulled out easily. The disease can be spoiled stem long distance by the presence of fallen shoots.

Management:

- Destruction of diseased clumps.
- Providing proper drainage.
- Changing the nursery site.

Azhukal disease / capsule rot / fruit rot – (*Phytoplithora parasitica car. Nicotincanae, Phytophthora palmivora, P.meadii, Pythium vexans*).

Symptoms:

Symptoms can appear on tender and matured leaves. Large circular, irregular, water soaked spots with black colour appear on leaves. The exposed portion of the unopened leaves my rot. Grey patches of irregular spots with brown margin are formed at the base of the leaf sleath. The basal portion rots and the pseudo stem break away at the collar region. The infection spreads to the underground plants and the rhizomes become rots. Small light brown lesions appear in the green tender fruits which fall of in 3-6 days leaving the small fruit stalk. The tip of inflorescence also rot.

Management:

- Removal and burning of infected paricles and rhizomes.

- Avoid moving of rhizomes sum diseased areas to healthy area for planting.
- Provide proper drainage.

Symptoms:

General chlorosis of young leaves – parallel streaks of pale green tissues running along the veins from midrib to the margins. Leaf sheath also shows stich stripes. In advanced stage the whole plant shows moxaic symptom clumps destoriates rapidly and plants are started. Rhizome shrivels and plants dies. If young clumps are attached they die before flowering.

Management:

- Collection and removal of infected clumps along with rhizomes and burning.
- Raising of nursery in diseases free areas.

Chenthal Disease/Leaf Blight (Colletotrichum Gloeosporioides)

Symptom:

Elongated, water soaked lesions of varying size appear on the upper surface of the leaf. The spots becomes brown to dark with pale yellow hole. Leaves wither and pseudo stem wilts. New shoots which develop are reduced in size. Flowers fail to develop. The inflorescence dry up stating from tip downwards. The affected gardens shows burnt appearance.

Management:

- Removal and destruction of affected leaves.

Nilgiris Necrosis (Nilgiris Necoros Virus)

Symptoms:

On leaves alternate light green to yellow patches or streaks appear parallel to the veins. A number of discontinues spocks appear on other areas of leaves. The virus spreads systemically in all the tillers. The leaves become small, stunted. The matme leaves tear off along the necrotic spots and shown shredded appearance leaves malformed.

Management:

- Use healthy rhizomes for planting.
- Rogue the infected plants.

Diseases of Ginger

Soft Rot or Rhizome Rot

Soft rot is the most destructive disease of ginger which results in total loss of affected clumps. The disease is soil-borne and is caused by *Pythium aphanidermatum. P. vexans and P. myriotylum* are also reported to be associated with the disease. The fungus multiplies with build up of soil moisture with the onset of south west monsoon.Younger sprouts are the most susceptible to the pathogen.

Symptoms:

- The infection starts at the collar region of the pseudostems and progresses upwards as well as downwards. The collar region of the affected pseudostem becomes water soaked and the rotting spreads to the rhizome resulting in soft rot.
- At a later stage root infection is also noticed.
- Foliar symptoms appear as light yellowing of the tips of lower leaves which gradually spreads to the leaf blades.
- In early stages of the disease, the middle portion of the leaves remain green while the margins become yellow. The yellowing spreads to all leaves of the plant from the lower region upwards and is followed by drooping, withering and drying of pseudostems.

Management:

- Cultural practices such as selection of well drained soils for planting is important for managing the disease, since stagnation of water predisposes the plant to infection.
- Seed rhizomes are to be selected from disease free gardens, since the disease is also seed borne.
- Application of *Trichoderma harzionum* along with neem cake @ 1kg/bed helps in preventing the disease.

Plantation Crops: Diseases of Areca Nut

Foot Rot or Anabe roga (Ganoderma lucidum)

Symptoms:

The leaf lets in the outer wholes of leaves become yellow and spreads to the whole leaf and the leaves drooping down covering the stem. Later, the inner whole leaves also become yellow. Subsequently all the leaves droop, dry up and fall off, leaving the stem alone. Then the stem also become bittle and easily broken by heavy wind. The base of the stem shows brown discoloration and oozing of dark fluid. Bracket shaped fustification of the fungus called 'anabe' appears at the base of the trunk. Roots become discoloured, brittle and dried. When infected trunk is cut brown discoloration can be seen up to one metre from ground level.

Management:

- Clean cultivation.
- Destruction of infected trees.
- Maintaining optimal plant population without over crowding.

- Providing good drainage facility and fertilizers and manures.

Yellow Leaf Disease (Phytoplama)

Symptom:

Yellowing of tips of leaflets in 2 or 3 leaves of outermost whorl. Brown necrotic streaks run parallel to veins in unfolded leaves. The yellowing extends to the middle of the lamina.Tips of the chlorotic leaves dry up. In advanced stage all the leaves become yellow. Finally the crown leaves fall off leaving of a bare trunk. Root tips turn black and gradually rot.

Management:

- Growing resistant varieties like true mangala and south kanara.
- Application of potassium and magnesium more than the recommended level.

Mahali/Kolerogo/Fruit Rot (Phyhophthora Arecae)

Symptoms:

Rotting and excessive shedding of immature nuts from the trees. The water soaked spots initially develop at the base of the nut. Fruit stalks and rachis of inflorescence are also affected. Nuts show large vacuoles and dark brown radial strands. Very often the top of the affected trees dries up resulting in withering of leaves and bunches. Affected nuts fall off and show the white mycelial growth of the fungus.

Management:

- Clean cultivation.
- Destruction of affected trees.
- Collection and burning of fallen nuts.

Bud Rot (Phytophthora Arecae)

The first symptom is the change of spindle leaf colour from green to yellow and then brown. The leaves rot and the growing bud rots causing death of the palm. The affected young leaf whorl can be easily pulled off. The outer leaves also become yellow and droop off one by one leaving a bare stem.

Management:

- Destruction and removal of dead palms and bunches affected by makali disease.

Flower Crops: Disease of Rose

Black Spot (Diplocarbon Rosae)

Symptoms:

- Black lesions with feathery margins surrounded by yellow tissue are found on the leaves. Infected leaves drop prematurely.

- Purple/red bumpy areas on first year canes may be evident.
- Plants may be weakened due to defoliation and reduced flower production may be observed.

Management:

- Use of Resistant Varieties.
- Cultural-Roses should be planted where the sun can quickly dry the night's dew.
- Space roses far enough apart for good air circulation Avoid overhead watering and keep foliage as dry as possible.
- Sanitation-Remove infected canes and burn diseased leaves.

Powdery Mildew (Sphaerotheca Pannosa)

Symptoms:

- The symptom appears as grayish-white powdery substance on the surfaces of young leaves, shoots and buds.

- Infected leaves may be distorted, and some leaf drop may occur.
- Flower buds may fail to open, and those that do may produce poor-quality flowers.
- It can occur almost anytime during the growing season when temperatures are mild (70 to 80 °F), and the relative humidity is high at night and low during the day.
- It is most severe in shady areas and during cooler periods.

Management:

- Collection and burning of fallen leaves.

Die Back (Diplodia Rosarum)

Symptoms:

- Drying of twigs from tip down wards.
- Blackening of the twigs.
- The disease spreads to root and causes complete killing of the plants.

Management:

- Pruning should be done so that lesions on the young shoots will be eliminated. Apply chaubatia pastic in the pruned area.

Rust (Phragmidum Mucronatum)

Symptoms:

- Damage to lemon yellow pustules appear on lower surface of the leaves and stems. Then the colour changes to blackish red.
- The affected leaves turn yellow deformed and fall prematurely.
- Die back symptom also appear due to weakening of the plant.

Management:

- Collection and burning of fallen leaves.

Disease of Jasmine

Phyllody

Symptoms:

Leaves become small malformed and bushy. In the place of flowers green leaf like malformed flowers are formed.

Management:

- Selection of cuttings from healthy plants.

Disease of Chrysanthemum

Rust (*Puccinia crysanthemi*) Rust is a serious disease especially in the early spring.

- The disease symptoms are in the form of brown blister-like swellings, which appear on the undersides of leaves.
- These burst open releasing masses of brown, powdery spores. Severely infected plants become very weak and fail to bloom properly.

Management:

- Early removal of infected leaves/plants helps to prevent the further spread of the disease.

Septoria Leaf Spot (Sepotria Chrysanthemella)

Leaf spots occur during cool-wet periods of the rainy season.

- Since the pathogens are spread through rain splashes the lowermost leaves get infected first.
- Serious infection may result in premature withering of the leaves; the dead leaves hang to the stem for some time.
- When flowering starts the infection occurs on flower buds, which rot completely.

Management:

- Destruction of disease debris and avoiding excessive irrigation is recommended.

Powdery Mildew (Oidium Chrysanthemi)

Infection is more severe in older plants under humid conditions.

- The growth of the fungus on the leaves appears as powdery coating. Infected leaves turn yellow and dry out.
- Infected plants remains stunted and fail to flower.

Management:

- Good ventilation and proper spacing for free circulation of air is recommended.

Diseases of Carnation

Fusarium Wilt (Fusarium Oxysporum F.sp. Dianthi)

In young plants, the first sign of the disease is fading or graying of the normal colour of the leaves with wilting of the leaves and young stems. It is followed by eventual collapse of the whole plant. When older plants are infected, similar symptoms are produced but the older leaves may show chlorosis followed by an indistinct purple-red discoloration. The vascular tissues of infected stems is stained dark brown. Mature plants show wilt symptoms over a period of several months before they die and eventually become straw coloured.

Fungus

The diseased plants should be removed immediately after noticing the disease. Complete root system and surrounding soil should be dug out and disposed off carefully. Soil solarization using clear transparent polyethylene film (0.1 mm thick) for 30 days gives satisfactory control. Grafting of susceptible cultivars like A lice, Fulvio Rosa, Gus Royalette and Johy, on to resistant rootstocks i.e. Arancio 25D, Exquisite, Heidi and May Britt and growing in soil naturally infested with fungus was also found to reduce the incidence of disease.

Alternaria Leaf Spot (Alternaria Dianthi)

The chief symptom is blight or rot at leaf bases and around nodes, which are girdled. Spots on leaves are ashy white. The centre of old spots are covered with dark brown to black fungal growth. Leaves may be constricted and twisted and the tip may be killed. Branches die-back at the girdled area and black crusts of conidia are formed on the cankers.

To reduce the disease incidence, humidity may be kept low by providing proper air circulation. Disease-free planting material should be used.

Bacterial Wilt (Burkholderia Caryophylli)

Symptoms:

The upper parts of established plants turn pale and wilt.The stem develops elongated discolored stripes and split open which is characteristic of the disease.The roots are rotted partially and the cortical tissues become sticky and shows discoloration, a tendency to straighten out instead of remaining curled. The leaves are twisted. The roots are generally lacking on one side and remain discolored. The base of the cutting is discolored, with an elongated brown area extending upwards. The bacteria invade xylem vessels and spread to infect the young shoots. The vessels are disrupted and the host plant produces meristematic tissues. This new tissue causes an uneven development of cortex which splits open longitudinally. Carnation wilt, a vascular disease may be induced by mechanical plugging, a toxin or a combination of factors. Carnation cuttings wilted when placed in a filtrate of a bacterial suspension of *B. caryophylli*. In the naturally infected carnation plants the xylem vessels are found partly plugged. *Lysis of xylem* vessels, is also observed. Infection is carried to upper portions of the branch into the leaves. Bacteria are not found in parenchyma tissues. Carnation cuttings suspended in bacterial filtrates in which proteins and other large molecular compounds are removed did not wilting symptoms.

Management:

Use of cuttings taken from upper parts of the healthy stock plants are less liable for infection and hence advocated as a control measure. Diseased plant debris should be collected and burnt. Overhead watering and splash watering should be avoided. Disease-free planting materials are to be used. Role cultivars viz., Elegance, Northland and Starlite are less susceptible to bacterial wilt.

Diseases of Gerbera (Gerbera asplenifolia, G. aurantiaca, G. Jamesonii (Barbeton Daisy), G. Kunzeane and G. Viridfolla)

1. Root rot (Rhizoctonia solani kuhn and pythium irregualsre): The infection result in stunted growth. Ultimately the entire plant dry. *Rhizoctonia solani* causes more losses and can attack older plants. Soil sterilization controls the diseases.

2. Foot rot and root rot (Phytophthora cryptogeas): The short stems blacken and rot. The leaves and flower die. Soil sterilization with vampam at 100 ml/square meter is very effective. Warming of soil (26 °C) reduces the incidence of the disease.

3. Blight/Grey mould (Botrytis cinera pers): The fungus kills young growing tissues. The flower

heads of the *Gerbera* growing in humid conditions show small, black spots on the ray florets. Deep planting, poor drainage and poor ventilation predispose the plants to infection. The disease can be reduced when the infected parts are removed and destroyed.

4. Powdery mildew (Erysiphe chichoracearum DC and oidium erysiphoides f.sp. gerbera): The fungus forms white powdery coating on the foliage. Spraying with wettabel sulphur controls *E. Chichoracearum*. Diseased leaves should be removed and destroyed.

5. Anthracnose (Colletotrichum gloeosporiodes penz): On *Gerbera jamesonii* Bolus ex. Hook.f. It has been reported from Karnataka and Maharashtra. The disease appears as circular, scattered, reddish brown spots. They coalesce with one another during moist weather involving large area and resulting in withering, rolling and drying of leaves. Excessive watering and crowding of plants should be avoided. Diseased leaves should be collected and burnt.

6. Blossom blight asnd stalk rot (Phytophythora palmivoras butler): It was reported from Maharashtra during 1971 and Karnataka. The disease appears as light brown, irregular, water soaked spots on flower stalks and petals. The spots increase rapidly and coalesce with one another and form distinct depressed lesions. Under humid conditions the infections become severe involving the entire flower head and resulting in blossom blight and stalk rot. The disease is favoured by drizzling rains and cool moist weather. The fungus is soil-borne and the infection starts from the base touching the soil. Use of disease free soil for cultivation reduces the disease incidence. Affected flowers should be collected and destroyed. Excessive watering should be avoided

7. Tobacco rattle (Tobacco rattle virus): Yellow or black annulated ring spots asre observed on the foliage. Soil steaming before each crop destroys the nematode vector, *Trichodorus* spp. and prevents the disease.

PERMISSIONS

All chapters in this book are published with permission under the Creative Commons Attribution Share Alike License or equivalent. Every chapter published in this book has been scrutinized by our experts. Their significance has been extensively debated. The topics covered herein carry significant information for a comprehensive understanding. They may even be implemented as practical applications or may be referred to as a beginning point for further studies.

We would like to thank the editorial team for lending their expertise to make the book truly unique. They have played a crucial role in the development of this book. Without their invaluable contributions this book wouldn't have been possible. They have made vital efforts to compile up to date information on the varied aspects of this subject to make this book a valuable addition to the collection of many professionals and students.

This book was conceptualized with the vision of imparting up-to-date and integrated information in this field. To ensure the same, a matchless editorial board was set up. Every individual on the board went through rigorous rounds of assessment to prove their worth. After which they invested a large part of their time researching and compiling the most relevant data for our readers.

The editorial board has been involved in producing this book since its inception. They have spent rigorous hours researching and exploring the diverse topics which have resulted in the successful publishing of this book. They have passed on their knowledge of decades through this book. To expedite this challenging task, the publisher supported the team at every step. A small team of assistant editors was also appointed to further simplify the editing procedure and attain best results for the readers.

Apart from the editorial board, the designing team has also invested a significant amount of their time in understanding the subject and creating the most relevant covers. They scrutinized every image to scout for the most suitable representation of the subject and create an appropriate cover for the book.

The publishing team has been an ardent support to the editorial, designing and production team. Their endless efforts to recruit the best for this project, has resulted in the accomplishment of this book. They are a veteran in the field of academics and their pool of knowledge is as vast as their experience in printing. Their expertise and guidance has proved useful at every step. Their uncompromising quality standards have made this book an exceptional effort. Their encouragement from time to time has been an inspiration for everyone.

The publisher and the editorial board hope that this book will prove to be a valuable piece of knowledge for students, practitioners and scholars across the globe.

INDEX

www.ingramcontent.com/pod-product-compliance
Lightning Source LLC
LaVergne TN
LVHW080657210926
848485LV00003B/390